# BLIND SPOT

How Industry Rescued America's
Great Depression Economy

## Jim Saunders

Copyright © 2020 by Jim Saunders. All Rights Reserved.

Copyright © 2020 by Jim Saunders

No part of this publication may be reproduced, distributed, or transmitted in any form or by any means, including photocopying, recording, or other electronic or mechanical methods, or by any information storage and retrieval system without the prior written permission of the publisher, except in the case of very brief quotations embodied in critical reviews and certain other noncommercial uses permitted by copyright law.

ISBN 978-1-7348624-0-9

Dedicated to
Patrick, Connor, and Aidan

# Contents

| | |
|---|---|
| List of Illustrations | vii |
| Introduction: The Deconstruction of Three Paths | 1 |

## Part 1 – Boom to Bust

| | | |
|---|---|---|
| Chapter 1 | The Roaring Twenties | 6 |
| Chapter 2 | A Great Depression Primer | 14 |
| Chapter 3 | The Great Puzzle | 25 |
| Chapter 4 | Building the World of Tomorrow | 31 |
| Chapter 5 | A Fresh Perspective | 40 |

## Part 2 – From Shop to Factory

| | | |
|---|---|---|
| Chapter 6 | The Robust Colonial Economy | 44 |
| Chapter 7 | America's First Factories | 47 |
| Chapter 8 | The American Industrial Revolution | 52 |
| Chapter 9 | The Rise of Big Business | 58 |

## Part 3 – A New Century with New Technologies

| | | |
|---|---|---|
| Chapter 10 | Launching Innovation | 63 |
| Chapter 11 | Henry Ford's Model T | 65 |
| Chapter 12 | Man's Dream of Powered Flight | 73 |
| Chapter 13 | Mechanized Farming | 77 |
| Chapter 14 | Freedom through Communications | 80 |
| Chapter 15 | The Electrical Power Grid | 83 |
| Chapter 16 | Electric Motors & Ball Bearings | 87 |
| Chapter 17 | Precision Grinding | 90 |
| Chapter 18 | The Evolution of the Modern Factory | 92 |

## Part 4 – How to Break a Robust Economy

| | | |
|---|---|---|
| Chapter 19 | Socialist Ideologies Come to America | 98 |
| Chapter 20 | Déjà Vu | 100 |
| Chapter 21 | Chaos in a Robust Economy | 103 |
| Chapter 22 | A Return to Mercantilism | 107 |
| Chapter 23 | Herbert Hoover the Wonder Boy | 120 |
| Chapter 24 | Smoot-Hawley Launches the Great Depression | 124 |
| Chapter 25 | FDR Moves into the White House | 130 |
| Chapter 26 | Financial Follies | 137 |
| Chapter 27 | Labor's Friend | 143 |
| Chapter 28 | Perpetual Jobs | 157 |

| | |
|---|---|
| Chapter 29  Farm Aid | 160 |
| Chapter 30  The Predator Attacks its Prey | 166 |
| Chapter 31  Punished for Excellence | 173 |

## Part 5 – America Turns Modern

| | |
|---|---|
| Chapter 32  The Nature of Invention | 181 |
| Chapter 33  Fire the Chauffeur | 186 |
| Chapter 34  From Pickups to the Big Rigs | 190 |
| Chapter 35  The Commercialization of Flight | 197 |
| Chapter 36  The End of the Cracker Barrel | 201 |
| Chapter 37  Larger, Smarter, and More Versatile Machine Tools | 206 |
| Chapter 38  Super Alloys | 211 |
| Chapter 39  Tungsten Carbide | 214 |
| Chapter 40  Stick Welding | 217 |
| Chapter 41  Inhaling Freon | 221 |
| Chapter 42  The Plastics Revolution | 224 |

## Part 6 – The Road to Recovery

| | |
|---|---|
| Chapter 43  World War Two Interrupts the Great Depression | 229 |
| Chapter 44  The Catalysts Driving the Recovery | 234 |
| Chapter 45  Peacetime Hangover | 246 |

## Part 7 – More Modern Marvels

| | |
|---|---|
| Chapter 46  The Time Machine | 253 |
| Chapter 47  The Mold in Dr. Florey's Coat | 256 |
| Chapter 48  The End of the Iconic Steam Engine | 261 |
| Chapter 49  Material Movers from Earth to Factory Floors | 268 |
| Chapter 50  I Should Just Go Home and Ride My Tractor | 271 |
| Chapter 51  Detroit Steel Back on Top | 273 |
| Chapter 52  From Shops to Labs | 276 |
| Chapter 53  The Post-War Housing Boom | 286 |
| Chapter 54  Homemakers' Helpers | 291 |

## Part 8 – Lessons Not Learned

| | |
|---|---|
| Chapter 55  Recovery by Millions of Know-Hows | 299 |
| Chapter 56  Questionable Lessons | 303 |
| Chapter 57  Measuring Grossly Distorted Productivity | 307 |
| Chapter 58  A Useless Appendage | 315 |
| Chapter 59  The Unemployment Paradox | 319 |
| Chapter 60  Another Great Depression? | 324 |
| Bibliography | 328 |
| Index | 339 |
| About the Author | 349 |

# List of Illustrations

**Chapter 2**

Photographer Dorothea Lange's famous photograph of Migrant Mother ... *Courtesy, Farm Security Administration-Office of War Information.* LoC C/N, 2017762891

Unemployment benefits aid begins ... *Courtesy, Farm Security Administration-Office of War Information.* LoC C/N, 2017872383 D

Red cross serving beverages to men ... *Courtesy, Farm Security Administration-Office of War Information.* LoC C/N, 2016879103

Bread line beside the Brooklyn Bridge ... *Courtesy, Farm Security Administration-Office of War Information.* LoC C/N, 2017872383

People living in miserable poverty ... *Courtesy, Farm Security Administration-Office of War Information.* LoC C/N, 2017763118

*Squatters along highway near Bakersfield* ... *Courtesy, Farm Security Administration-Office of War Information.* LoC C/N, 2017759223

**Chapter 4**

Part carnival side show-part valuable research ... *Courtesy, Manuscripts and Archives Division, The New York Public Library.* "Infant Incubator – Woman looking at babies in incubators." Image ID, 1675847

General Motors – Futurama ... *Courtesy, Manuscripts and Archives Division, New York Public Library.* "General Motors – Futurama – Visitors in moving chairs viewing exhibit." Image, ID 1674373

Pontiac transparent "ghost car" ... *Courtesy, Manuscripts and Archives Division, New York Public Library.* "General Motors – Crowd viewing transparent car at dedication." Image ID, 1674229

RCA Television ... *Courtesy, Manuscript and Archives Division, New York Public Library.* "Radio Corporation of America (RCA) – Harvey Gibson, Miss Television, and James E. Robert standing with television" Image ID, 1681015

Miss Chemistry ... *Courtesy, Hagley Museum and Library* ' *"Miss Chemistry models nylon stockings at the New York World's Fair."* Image ID, 1984259_121412_013

Elektro and Sparko ... *Courtesy, Manuscript and Archives Division, New York Public Library.* "Westinghouse – Mechanical Man and Dog (Elektro and Sparko)" Image ID 1686391

Theme Center ... *Courtesy, Manuscript and Archives Division, New York Public Library.*

"Theme Center – Trylon and Perisphere – View of Trylon and Perisphere with sculpture, fountains, man and woman in foreground" Image ID 1684531

## Chapter 7
Jedidiah Strutt 18th century cotton mill. *Courtesy, Wikimedia Commons scanned from 1819 Rees' Cyclopedia*

## Chapter 11
1913 Ford Model T ... *Courtesy, The National Museum of American History, Smithsonian Institution.* ID No. TR.311052

## Chapter 12
With Orville Wright at the controls. *Courtesy, Smithsonian Air and Space Museum, Smithsonian Institution.* ID SI2003-3463
U. S. Army Air Mail Curtiss JN-4HM ... *Courtesy, Smithsonian Air and Space Museum, Smithsonian Institution.* ID 2000-6150
White I-30 Tractor Pulling Ford Tri-Motor Airplane ... *Courtesy, Wisconsin Historical Society.* image ID 7213

## Chapter 13
Doctor Onsgard harvesting ... *Courtesy, Fillmore County Historical society*, https://reflections.mndigital.org /fch:189

## Chapter 16
Ball Bearing ... *Courtesy, Grayhawk Press*
Tapered roller bearing ... *Courtesy, Grayhawk Press*

## Chapter 18
Machine shop classroom ... *Courtesy, Wisconsin Historical Society.* image ID 7952

## Chapter 27
Armory, Columbus, Ohio ... *Courtesy Library of Congress – Detroit Publishing Company.* 2016803777

## Chapter 33
General Motors Turret Top Promotion. *GM Media Archives.* Image No. 211248
1939 Ford DeLuxe ... *Courtesy, The Henry Ford.* ID THF90290
Transportation to Fair – Women with car and trailer. *Courtesy, Manuscript and Archives Division, New York Public Library.* Image ID 1685005

## Chapter 34
International 1939 D-Line Truck *Courtesy, Wisconsin Historical Society.* image ID 122108
Semi-tractor with 5th wheel-1939. *Courtesy, Farm Security Administration – Office of War Information.* LoC C/N 2017719135
International D-300 Truck … *Courtesy, Wisconsin Historical Society.* image ID 121289
Greyhound Lines bus-1937. *Courtesy, Library of Congress https://lccn.loc.gov/2016878403*
Pre-war Kenworth Fire Truck. MOHAI, *Courtesy, Museum of History and Industry.* 1983.10.14407.3
Milkman in front of Divco … MOHAI, *Courtesy, Museum of History and Industry.* 1983.10.13902.1
A driver sits in an International truck TiltTop Bottle Capper-1929. *Courtesy, Wisconsin Historical Society.* image ID 49202

## Chapter 35
A DC-3 offshore at Tigvariak Island, Alaska. *Courtesy, US National Oceanic and Atmospheric Administration*
Elevated view of group of men and women boarding an American Airlines … *Courtesy, Wisconsin Historical Society* Image ID:62924

## Chapter 36
Empty cans are filled with peas … *Courtesy, Farm Security Administration-Office of War Information.* LoC C/N 2017780255/
A ten-arm Owens automatic bottle machine, ca. 1913. *Courtesy, Lewis Hine*
Filling milk bottles at creamery … *Courtesy, Farm Security Administration-Office of War Information.* LoC C/N 2017785014
Placing packaged goods on display rack … *Courtesy, Farm Security Administration-Office of War Information.* LoC C/N 2017785132

## Chapter 40
A "Wendy Welder" … *Courtesy, U.S. Office of War Information.* LoC C/N 2017697447

## Chapter 48
GM 1947 Electromotive F-Series Astra Liner … *Courtesy, General Motors Archives.* ID Number 166158

## Chapter 49
Bulldozer used in grading during the construction *Courtesy, Farm Security Administration-Office of War Information.* LoC C/N, 2017871713

Laying sidewalks with grader at New York World's Fair-1939. *Courtesy, Manuscript and Archives Division, New York Public Library.* Image ID 1675481

## Chapter 50
1946 catalog illustration of an International Harvester Farmall HV tractor. *Courtesy, Wisconsin Historical Society.* Image ID: 28761

## Chapter 51
Ford Model A ... *Courtesy, The Henry Ford.* ID THF120026
1949 Ford Tudor sedan ... *Courtesy, The Henry Ford.* ID THF90472

## Chapter 52
O-ring. *Courtesy, Grayhawk Press*

## Chapter 54
A window display at the Wisconsin Power ... *Courtesy, Wisconsin Historical Society.* Image ID 73462
Testing a Westinghouse refrigerator-1937. *Courtesy, Wisconsin Historical Society* Image ID. 14963
1947 International Harvester refrigerator. *Courtesy, Wisconsin Historical Society* Image ID. 12017
Saleswoman demonstrating a Maytag washing machine to a customer (1937). *Courtesy, Wisconsin Historical Society.* Image ID 14889
The Bendix Home Laundry was a homemaker's dream ... *Courtesy, MOHAI Museum of History and Industry.* 1983.10.15984.2

# Introduction: The Deconstruction of Three Paths

The Great Depression ranks as one of American history's most scrutinized events. Only the Civil War and World War II hold more historical interest than this 1930s catastrophe. 90 years later, its story has all the elements of a classic thriller. For more than a decade, the tale of America's infamous economic panic provided a diverse collection of hardworking underdogs, flamboyant protagonists, and loathed antagonists. The plot contains several dramatic twists, while the tremendous recovery led to an extraordinary feel-good ending. Yet, two controversial mysteries remain unresolved: What caused the Great Depression? Did World War II spending drive the recovery?

Mainstream scholarship suggests that academia has a lock on those answers: a cataclysmic failure of capitalism caused the Great Depression; the New Deal offered some relief; and finally, massive World War II military spending rejuvenated the still-sluggish economy.

Yet, the debate trudges on and at times intensifies as disenfranchised academics and enthusiasts challenge the establishment's view of the *panic* years.

One indisputable Great Depression truth is the scale of the economic damage endured by Americans during the 1930s and into the 1940s. The unemployment rate between 1931 and 1940 peaked at 25% and never dropped much below 15%. In the first few years, a third of America's banks closed. When the economy struck bottom, many of the nation's business sectors' revenues dropped by 50%. Then came the war. Nearly 400,000 Americans died in combat during World War II, with total casualties exceeding one million. Back home, civilians lived with widespread shortages and harsh rationing. Even the post-war recovery turned into a multi-year struggle.

Another undeniable truism is the enormous impact the Great Depression has had on America's social-political system. The policies that emerged from the Great Depression influenced the character of our present-day political philosophy more than any other era since the Civil War. Consider a few examples:

- During the post-World War I isolationist era, a nation grown weary of fighting Europe's latest battle cut military expenditures to 3% of the gross domestic product (GDP). After World War II ended, propped-up defense spending intended to support the world's number-one military power consumed a burdensome 10% of the nation's finances.

- Republicans held control of all three federal branches throughout the worst years of the Great Depression (1929-1933). Conversely, the Democrats owned those majorities for the bulk of the recovery period.
- In less than four years, the nation moved from the most liberty-oriented government on earth, the laissez-faire model, to one run by political progressives chasing Europe's socialist-leaning agenda.
- The federal government took over most of the entitlement and safety-net programs, once the responsibility of churches, private charities, and local municipalities.
- The US Treasury began its transition from a gold to a fiat standard.
- Organized labor, instead of operating on the fringes of America's workforce, matured enough to become an integral part of the national economy.
- Republicans in 1921 and 1930 passed trade bills with excessive tariffs; Democrats, during the late 1930s, broke down those protectionist barriers through international free-trade agreements.
- The failed social experiment of national alcohol prohibition enacted in 1919 ended in 1933.

These examples represent profound changes in American culture and are well documented by a battery of scholars. Notice that all these topics only center on government and politics. Dig deeper into the inner workings of the Great Depression and a whole other dynamic emerges, one that is more dramatic than the intense personal hardships, the sweeping political conversion to a socialist-leaning agenda, or the hard-fought economic debates.

The most spectacular outcome from the Great Depression is that while unemployment soared and social leaders dove into damage-control mode to fix an ailing economy, technology modernized the nation. Across dozens of major industries, in a short time span, a surge in new technologies transformed American culture. Compare a 1929 automobile, refrigerator, washing machine, bus, or locomotive to those built in 1939 or to the post-war models of the late 1940s. In each case, the latter designs bear a noticeably closer resemblance to their twenty-first century counterparts than their 1920s predecessors. There are countless other excellent examples, not just in transportation and home appliances, but in pharmaceuticals, agriculture, home construction, material sciences, packaging, industrial processes, and manufacturing equipment.

Despite the tremendous size and scope of America's 1930s industrial economy, influential historians snub the phenomenon. The striking omission emphasizes academia's blind spot toward the wonders of capitalism.

If mainstream academia lacked insight into the early midcentury industrial revolution, then what else did they miss? As it turns out, quite a bit. For one, how did an abundance of new technological marvels coexist with both record high unemployment and a dismal business climate?

The Great Depression whodunit storyboard needs a manufacturing perspective to answer the previous question and others, such as:

- Which catastrophes reversed the prosperity of the booming Roaring Twenties?
- How revolutionary were those 1930s game-changing technologies?
- How did establishment politicians and academics harm the economy with their politically charged battles against capitalism?
- What role did World War II play in the recovery?

And one more huge issue for the history sleuth:

- Why can't scholars' sophisticated post-war productivity metrics gauge the true economic benefits of modern technology?

To gain insights into its conflicting nature, this book investigates three Great Depression topics:

## America's Modernization

The number-one takeaway from the Great Depression should be that despite the economic turmoil, America turned modern. Well beyond a collection of scientific discoveries, the panic years' industrial revolution required enormous amounts of innovation, automation, and expansion leading to unprecedented productivity gains.

## The Recession's Paradox

How did so much despair go hand in hand with an expanding factory economy? For starters, social reformers attacked a problem that did not exist. They envisioned a gentler avant-garde version of capitalism. After all, technical revolutions have their darker side. With technological change came chaos. For example, corporate consolidation caused weaker businesses to either adapt, merge, or fail. New technologies forced many workers to learn new skills as traditional trades became obsolete. The marketplace turned more complex. Some communities expanded as others shrank.

This transformation sent distress signals to a growing class of progressives, which included left-leaning academics, bureaucrats, social advocates, and politicians. With a core belief that capitalism contained a destructive element, their knee-jerk reaction to the Stock Market Crash of 1929 was to check America's rebel industrial movement with regulation and taxation. This caused more than its fair share of unintended consequences, which delayed what should have been a quick recovery. Then came the New Deal (1933–1937) with its horde of detrimental policies. Later, as advanced capitalism collided with the weight of an expanding federal government a conflicting combination of widespread poverty and modernization emerged.

**Easy Money**

Neither academia nor the Federal Reserve (Fed) have been able to measure commercial growth. Instead, to explain the catastrophic downfall and the amazing recovery, financiers built a parallel, virtual economy based on money as a growth enabler. This adoption of Keynesian principals, named after John Maynard Keynes (1883-1946), led to newly printed money in modest doses giving the illusion of an ideal growth rate. In reality, their "convenient" model treats America's industrial complex as a simple math problem, which elevates the clout of progressive ideals yet diminishes the power of capitalism. That same financial system continues to plague America's twenty-first century economy.

✪✪✪

Great Depression history deserves a deconstruction; one from an industrial perspective. Chapter 1 begins the journey with an overview of the 1920s. The Roaring Twenties earned its moniker because of the decade's economic robustness and cultural vibrancy, yet mainstream academia characterizes the decade as lacking enough substance to sustain its hyper growth rate. Just how extraordinary was the American economy when it crashed in 1929?

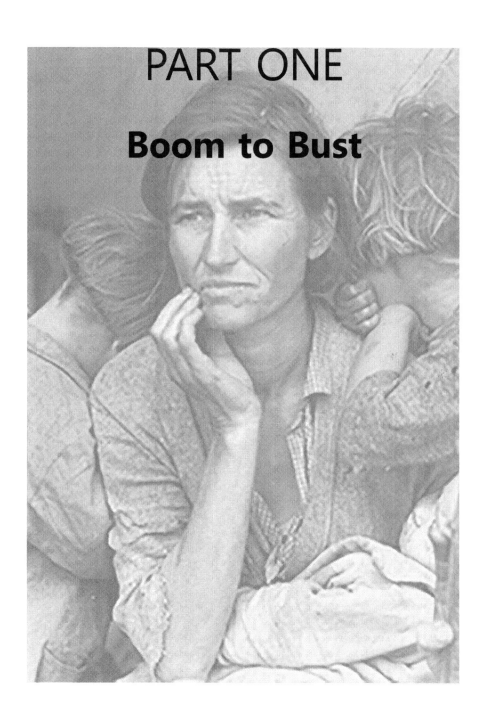

# PART ONE

# Boom to Bust

# 1
# The Roaring Twenties

To appreciate the full significance of the Great Depression is to realize how far the economy fell from its pinnacle, reached in September 1929. That lofty peak concluded the Roaring Twenties. Much more than a strong economy with low unemployment, the era's technical and cultural intensity dwarfed those of any earlier time in world history.

A considerable number of technologies set up a Roaring Twenties economy. Four game-changing innovations led the charge: the electrical power grid, the internal combustion engine, electronics, and the motorized factory. Combined, they brought America another step closer to becoming modern. Sustained by those engineering marvels, the floodgates of innovation burst open.

In 1900, just 3% of Americans had their homes wired for electricity, but by 1920, the ever-expanding electrical grid powered 35% of American households. Over that same time frame, the automobile business graduated from backyard shops that built rare novelties into a leading industry with over one million cars roaming America's roadways. Wireless telegraphs, and soon the introduction of radios, brought a communication revolution that would rival Gutenberg's movable-type print shop. In 1903, the Wright Brothers reached a major milestone toward humankind's dream to master flight. Communication and air travel combined to reinforce the metaphor that technology made the earth smaller. Inventors also pioneered a slew of innovative, labor-saving devices. As a result, craft industries transformed one by one as they switched over to some degree of mass-production. The outcome? More affordable and higher-quality goods became available to Americans of all income classes.

✧✧✧

All that inventive hustle and bustle spurred a vigorous economy—that is, until 1917 when Europe drew the United States into the escalating global conflict. An overwhelming commitment to the war effort brought a momentary halt to the phenomenal growth rate of American affluence.

After the war, the economy gained back some of its economic momentum. The nation appeared ready to continue where it had left off in 1917. Unfortunately, consumer optimism turned out to be a false start as the economy could not escape the traditional post-warfare hangover. In January 1920, fourteen months after the allies celebrated Armistice Day, a sharp recession slammed the United States. At its worst, eight months into the slide, a lagging business climate caused the unemployment rate to double and the stock market to fall by 40%. Fortunately, the deep post-war downturn ran a natural course on the road to recovery and did not last long. By June 1921, the recession had ended. The federal government's hands-off approach to the economic downturn continues to be a valuable history lesson.

Back on track, the world's number-one industrial superpower resumed its journey toward modernization, and along the way created one of the most vibrant decades in history. The eight-plus-year streak lasted from the economic reboot in mid-year 1921 until the October 1929 stock market crash. The economy sizzled over that span, and industry once again increased its pace of modernization, gaining Americans access to a host of new consumer and industrial goods. The low unemployment rate, now estimated to be about 3% to 5%, placed the number of unemployed workers near the theoretical lower threshold, the definition of full employment. A strict immigration policy limited the turmoil of too many low-wage workers chasing the few low-skilled jobs available.

A cautious stock market took a couple more months to catch up to the post-World War I recovery already in progress. The Dow Jones Industrial Average (DJIA) had bottomed out at 64 points in August 1921. The stock market then turned its sites upwards, maintaining an extended, steady growth rate that lasted until the summer of 1928. Then over the next year, a frenzied wave of speculation further energized the market. The surge of optimism accelerated the already rising stock market, which drove the Dow to a historical high of 381 points on September 3, 1929. The 381 mark represented a growth factor of over sixfold since the start of the 1921 post-war recovery.

Most Americans could not afford to dabble in stocks; however, the majority who did earned handsome returns. For those who invested heavily, the odds were that they became extremely wealthy. Throughout the 1920s, the size of the United States middle class swelled, but the number of the rich exploded. In 1921, the US had 21 millionaires, by 1927, the exclusive club counted 15,000 members. Furthermore, this impressive accomplishment occurred during a time of a relatively tightened monetary policy by the Federal Reserve.

The Roaring Twenties also left a legacy of conflict. While obscene opulence flourished, in stark contrast, millions struggled with poverty. Workers, even those with steady work in factories, mills, or on the farms, often found their meager wages meant a constant battle for them to make ends meet. For the less-fortunate, hardships such as old age, disabilities, and language barriers often made for a difficult life. In Manhattan, African American writers and artists contributed to the vibrant 1920s culture with the Harlem Renaissance, yet across the nation severe discrimination kept many of the 11 million African Americans well below the poverty line.

Despite these trying social issues, the ongoing transformation of Western civilization improved the quality of life for Americans. Labor-saving machines lessened the drudgery and hazardous conditions of a hard day's work. The poor had more opportunities to increase their purchasing power thanks to a new age of manufactured goods previously affordable only to the rich, if even available. Overall, the decade had such richness and depth that it offered something modernistic for everybody—especially in entertainment.

The 1920s marked an era in which Americans strengthened their love affair with sports. Fans flocked to athletic venues, leading to the construction of over 70 large stadiums by colleges, municipalities, and sports teams. The popularity of baseball and football funded such notable outdoor arenas as Yankee Stadium (capacity 82,000), the Rose Bowl (capacity 80,000), Soldier Field (capacity 100,000), and the Los Angeles Coliseum (capacity 75,000). Nationwide, a fitness trend financed the building of countless basketball courts, gymnasiums, and swimming pools.

Sports stars became national legends. Their fame endures nearly a century later. Jack Dempsey (1895–1983) reigned as boxing's Heavy Weight Champion of the World for most of the decade. Golfer Bobby Jones (1902–1971) dominated the US Open. Football coach Knute Rockne (1888–1931) guided the Notre Dame Fighting Irish to compete in three collegiate national championship games. Baseball's Boston Red Sox sold Babe Ruth (1895–1948), one of baseball's best left-handed pitchers, to the New York Yankees in 1920. The Yankees switched Ruth to right field, and the Bambino became the game's greatest all-time slugger. The Babe won the American League home run title 8 out of 10 seasons during the 1920s.

Americans also fell in love with movies. From the small towns to the big cities, people of all ages enjoyed an afternoon matinee or a night out to see a movie. Moviegoers took in a film at their local cinema an average of once a week. The stars of Hollywood, home of the major studios since 1912, became so famous that they demanded huge annual five-digit salaries, an impressive figure for the fledgling industry. Charlie Chaplin (1889–1977), already the world's most well-known celebrity by 1920, remained popular throughout the decade by playing his signature character, the Little Tramp. In 1926, a small studio named Warner Brothers released the motion picture *Don Juan*. The hit movie was the first to use a Vitaphone, which employed a disc to play recorded background music synchronized with the film. The next year, 1927, Warner

Brothers introduced talkies with *The Jazz Singer* starring Al Jolson (1886–1950). The profitable film again used a Vitaphone, but with more precision the process accurately synchronized the music with a singer's moving lips.

For home entertainment, people listened to the radio or played phonograph records. The first AM radio broadcasts began in the early 1920s. By decade's end, many American families owned at least one radio. Even during those early days, studios broadcasted a variety of programs such as news, sporting events, serial programs, and of course music. On phonograph discs, music lovers had an eclectic range to choose from including popular, jazz, blues, ragtime, Broadway, country, symphony, chamber, string quartet, and opera.

On January 17, 1920, the controversial Eighteenth Amendment, which outlawed the sale of alcohol, went into effect. From then on, until its repeal in 1934 (the Twenty-First Amendment), the selling and consumption of liquor went underground. Prohibition led speakeasies, often run by organized crime families, to proliferate. One of the nation's most infamous gangsters was Chicago's Al Capone (1899–1947), who employed a staggering 1,000 hired guns. His and similar organizations of corruption also lured many policemen and politicians onto their payrolls. Symbolized by fast cars and Tommy guns, the competition among criminals to sell alcohol resulted in unprecedented violence. The national murder rate increased 40% during Prohibition and then dropped 40% after it ended—an indication of the cost of depriving Americans of liquor.

Women made great strides on closing the inequality gap with men. The Nineteenth Amendment, signed in 1920, gave women the constitutional right to vote. However, women's liberation reached well beyond politics, especially for many of the youthful generation. Edgy young ladies known as flappers smoked cigarettes and sported short hair. They wore wispy dresses cut at the knees, long beaded necklaces, sheer stockings, and lots of makeup. Youngsters danced to a new racy dance craze, the Charleston. Not content to be just homemakers, women entered the workforce in greater numbers, armed with a newfound sense of independence and confidence.

America's extraordinary path toward modernization had much to do with the 1920s cultural revolution. With technological advancements came more leisure, better job opportunities, improved health, and overall, a higher quality of life. The technology explosion allowed not just the rich but also the middle and lower classes access to advantages and luxuries mostly unobtainable by earlier generations. Consider the range and depth of a few of those advancements.

By 1929, a more expansive electrical power grid brought enormous amounts of affordable power to 90% of Americans. Quiet, odorless, and relatively inexpensive, electricity delivered to each household the equivalent power of a modest-sized, early nineteenth-century steam engine. First built to power light bulbs, electricity's essential benefit emerged with the development of domestic electrical appliances. Efficient motorized factories soon followed. The well-planned buildings occupied a single level instead of the traditional four or five-story mill-styled designs.

The nation's first housing boom both chased and contributed to urban sprawl. Americans in record numbers abandoned tenement housing for affordable, single-dwelling homes often located outside the city limits. Architectural styles included colonials, New England Cape Cods, and California bungalows. Sears & Roebuck sold prefabricated homes for buyers on lower budgets. The most modern of those homes featured comfortable living with convenient electricity, running hot and cold water, accessibility to a public sewer system, and central heat. A new generation of electric-powered, labor-saving devices eased the drudgery of household chores.

The electrical grid lit up America's magnificent cities, which, as the population grew larger and wealthier, began to spread both out and up. Steel buildings aided by speedy, automatic electric elevators raised the skylines into the clouds, particularly in Chicago and New York. At the end of the decade, builders laid the foundation for two of the world's most renowned modern structures: the mighty Empire State Building and the stylish art deco Chrysler building.

To extend metropolitan areas hampered by waterways, American cities funded the construction of massive steel-suspension bridges. From the 1920s, four new bridges including the Ambassador (Detroit), Delaware River (Philadelphia), Bear Mountain (Hudson River), and Williamsburg (New York City), joined the forty-year-old Brooklyn Bridge as the world's five longest suspension bridges. Crews began construction on the magnificent George Washington Bridge (GWB) in 1927, and it opened in 1931. The GWB doubled the span of the previous longest suspension bridge at over 3,500 feet.

Technology, with the help of affordable compact farm tractors, trucks, electric-powered irrigation, electric lighting, and chemical fertilizers continued to revolutionize farming. Mechanization gave farmers the tools to produce more goods per acre of land. Low-cost farm products fed and clothed an energetic population, and as the largest agricultural-goods producer in the world, the valuable surplus of farm goods accounted for nearly half the worth of American exports. Modern farming reduced the need for lots of farmhands and millions of work horses, which freed up additional farmland previously used to feed the horses.

In Detroit, the automobile industry became America's fastest growing and most iconic sector. A generation earlier, the sight of an automobile rumbling by brought reactions of amazement and fright. By the 1920s, America's new big industry flooded the nation's highways with rolling steel backed by lots of horsepower. Car sales throughout the latter 1920s hovered between three and four million vehicles per year.

To accommodate an automobile-crazed America on the go, states built multi-lane, paved motorways that crisscrossed the nation. Those highways linked population centers to the countryside and vacation destinations. The Lincoln Highway guided motorists cross country from New York City to San Francisco. New York's Taconic State Parkway networked city dwellers to state parks via a tree-lined motorway. By 1925, highway US 1 formed a near-continuous route from Bangor, Maine, to Key West, Florida, a distance of 2,000 miles. Along those routes sprang service stations, restaurants, and motor hotels. The term motel joined an expanding American lexicon inspired by new technologies.

Larger, more powerful planes led to the growth of commercial airlines. A few years earlier, World War I biplanes were the most advanced flying machines of the time. Their lower-horsepower engines required minimum weight, thus the fuselage and wings were fashioned from wood and cloth and bound by steel wires. By the late 1920s, tri-engine, aluminum-clad monoplanes (fixed-wing aircraft) signaled a new era in aviation. Thanks to the continued development of flight, over 100 commercial carriers offered scheduled passenger and mail services between dozens of airports. Airlines augmented an already extensive railroad system and the expanding road network.

Electronics, along with aviation and the material sciences, developed into one of the more popular engineering disciplines. A wide variety of do-it-yourself kits allowed electronic hobbyists to build AM or short-wave radios. That kind of tinkering encouraged many young men to enter engineering fields. In the case of electrical engineering, the subject eventually split into two disciplines: power and control. The first focused on pumping up the nation's electrical grid with unfathomable amounts of efficient energy. The second aimed to finesse the flow of electrons into performing fantastic magic by the transmission of sound either through copper wires or the atmosphere.

Telephone technology became more complex as the number of local and long-distance calls swelled. Due to the increase in phone traffic, Bell Telephone (aka Ma Bell) struggled to offer enough capacity for its market. The telephone industry's evolving research and development culture eventually formalized their challenges. In 1925, Western Electric joined forces with Bell to build Bell Laboratories, an early version of Silicon Valley, which became America's premier research facility, a spot they occupied for the next few decades.

Medicine continued its rapid twentieth century transformation. Pharmaceutical companies supplanted the snake oil business while an informed public learned to avoid fake doctors. Hospitals played an expanding key role in health care. For both whites and minorities, infant mortality and maternal death rates dropped by over 20% during the 1920s Fatalities from diseases such as diphtheria and polio also plunged.

Hidden out of sight, behind the walls of cutting-edge factories, came an "invisible" industrial revolution. As manufacturers learned how to squeeze lots of horsepower out of a small electric motor, they abandoned the traditional rat's nest of shafts, belts, and pulleys, the standard method of power transfer employed by factories since the late eighteenth century. The new technology radicalized manufacturing versatility and efficiency. To some degree, every industry in the country benefited from those upgrades.

Labor efficiency across multiple sectors increased. Due to the high level of technology packed into and exiting from factories, American workers, whether on the manufacturing floor, in the office, on the road, in retail, or at the farm yielded more output for each hour of labor.

Commerce reigned on Main Street. A new era of chain department stores, supermarkets, automobile dealerships, appliance stores, and other businesses attracted an increasing number of customers. This new consumer era arrived with mixed concerns. Had increased materialism corrupted traditional social values? Was the tradeoff worth the additional commerce stoking the economy? Americans combined their greater purchasing power with a desire to buy the latest and greatest technological marvels. Increased corporate income provided funds for development, which created more gadgets, leading to more spending. The greater revenues then encouraged the invention of more gadgets. There seemed to be no end to the capitalist cycle that kept American consumers both upbeat and thriving.

Nevertheless, the hyper financial sector developed some cracks by decade's end. For investors on Wall Street, the Roaring Twenties had been a great ride. Then by the autumn of 1929, a sputtering stock market signaled an end to the fantastic streak. On Monday, October 28, 1929, the DJIA fell 38.4 points, a 13% drop. The next day, the infamous Black Tuesday, it fell another 30.5 points, a 12% drop. In total, the stock market lost 33% of its value over the final two weeks of October 1929.

As the 1920s ended, 8 years after the last major recession, the developing bear attitude in the stock market sent ripples throughout the entire economy. Across multiple sectors, a declining trend in consumer confidence matched the doom and gloom attitude infecting stocks. The economic panic led to deteriorating commerce and sporadic layoffs, bankruptcies, and bank closures. All of which induced more of the same. The economy had come full circle as decline replaced prosperity.

Americans at the time did not realize the severity of the downturn, and for good reason. Judged by the experiences of previous recessions, the crisis should have amounted to nothing more than the correction of an over-exuberant market. Instead, the recession evolved into one of the worst economic panics in American history.

The Great Depression was on.

# 2
# A Great Depression Primer

The Roaring Twenties left few clues that the thriving interval would end in economic turmoil. Despite the decade's reputation for prosperity, the period had its share of troubles. Distressed hot spots included: regional housing busts, particularly in Florida; farmers struggling to survive in an era of plunging prices for agricultural goods; a wave of bank failures; and lower-class frustration over a real, or at least perceived, expanding income gap. Those events, while devastating to some, accounted for mere background noise in an otherwise booming economy. That is, until panic struck in late October 1929.

The trigger came from the hub of Roaring Twenties opulence, the stock market. As stocks began to flatten during the autumn of 1929, Wall Street sent mixed warning signals on the economy's health. Investors and analysts had a "which way is the wind blowing" outlook toward the fitness of the stock market. When stock prices dropped, analysts warned that the market was over-valued due to reckless speculation and needed a downward correction. When the market reversed amid renewed optimism, analysts applauded the robustness of stocks and interpreted those increases as proof of America's strong economic core.

In hindsight, the crash appears inevitable. Investor exuberance had driven the stock market to unsustainable heights—a bubble. The bubble burst amongst record trading and losses. The first casualties were the financiers who had bought stocks on margins with borrowed money. Some lost everything. Disgraced and financially ruined, legend has it that brokers began flying out of upper-story windows. Today, scholars consider the suicides more hyperbole than fact. However, stories such as the desk clerk of a hotel near Wall Street who asked guests if they wanted a room for sleeping or jumping does fit the mood of the times.

An additional wave of casualties included those who had much of their wealth invested in the stock market and optimistically left their investments on the sinking ship too long. Only those who got out quickly could escape with their fortunes intact. Groucho Marx (1890–1977) did not sell in time. According to Maury Klein (b. 1939), author of *Rainbow's End* (2003),the comedian once declared, "I only lost $240,000—because that's all I had." Groucho was lucky. Still in his prime, he hit the concert trail and rebuilt his fortune. Billy Durant (1861–1947), the famed founder of General Motors and a major stock market player, lost almost everything and never recovered. He ended his career flipping hamburgers at his bowling alley.

As the stock market continued to tumble, it dragged down the rest of the economy. The domino effect caused production rates to fall and the unemployment rate to rise. In retrospect, the stock market acted as no more than a catalyst. More powerful forces drove the economy into a much deeper state of despair. In 1934, a British economist, Lionel Robbins (1898–1984) published a book on the recession, entitled *The Great Depression*. The book's title stuck as an apt name for the ongoing economic catastrophe. By the mid-1930s, the term "The Great Depression" had become a part of the American vocabulary.

The Great Depression earned its reputation as America's worst twentieth-century economic disaster for several reasons. It exceeded all others in depth, scope, and duration. The recession's severity caused suffering for millions of people, many of whom endured economic hardships for several years. The poorest of them often had to worry about necessities such as adequate shelter and where their next meal might come from. These people were not merely between jobs, they became chronically unemployed and desperate, dependent on friends, family, and as a last resort, charity. Not just a regional conflict, the Great Depression brought widespread poverty to nearly the entire country as it slowed down practically every commercial sector. The panic forced the majority of businesses to shrink payrolls, but the sinking economy inflicted its harshest consequences on rural communities and farmers. Southern sharecroppers and victims of the infamous Dust Bowl crisis suffered the most. Farm goods prices had fallen during the previous decade, but this time a weakened manufacturing sector lost the ability it once had to hire many of the displaced agricultural workers.

Another legacy left by the Great Depression was President Roosevelt's (1882–1945) stimulus initiative known as the New Deal. A continuous flow of three to four million workers cycled through the New Deal's public works programs, mainly the Civilian Conservation Corps (CCC) and the Works Progress Administration (WPA). Today, many of those 80-plus-year-old hiking trails, buildings, tunnels, and bridges serve as reminders of the massive effort taken by the federal government to put Americans back to work.

In sharp contrast to the vibrant culture of the Roaring Twenties, the hard economic times of the 1930s took on a much more somber tone. In literature, John Steinbeck's (1902–1968) bestseller *The Grapes of Wrath* (1939) fits that mood. In music, the pop tune *Brother, Can You Spare a Dime?* echoed the feelings of the underclass working man.

Since America was the world's largest importer and exporter, the economic crisis had a significant global influence. When the powerful American economy crashed, it took most of the industrialized world with it. Still reeling from the massive debt of World War I, the struggling European economy enabled a disheveled Germany and a fascist Italy to take their command economies to the frightening next level, plunging Europe into another melee which became World War II. The truce period between the great wars lasted but a single generation.

By 1933, the dire economic conditions had swept in a new regime of federally elected officials and the nation moved from a heavy Republican majority to a heavy Democratic majority. The federal government's new leadership immediately introduced radical legislation to battle what they considered a desperate situation. Thus, the era of the Great Depression marked a dramatic, historic shift in political policy. The new guard reversed a 150-year stretch where politicians supported a minimalist federal government. Instead, the recovery effort popularized an activist central government, which embraced the New Deal and favored numerous elements of a command/cradle-to-grave economy. A significant percentage of those initiatives survived the recovery altering the American political ethos. Today, our modern political culture owes much of its provenance to the Great Depression.

In 1935, the Roosevelt administration established the Farm Security Administration (FSA). The agency became controversial due to its Soviet-styled mission of building collective farms on federal land, but it did produce one lasting benefit. The FSA hired professional, well-equipped photographers to capture on film the struggles of rural life during the Great Depression. Fortunately, many of those photographs found their way to accessible archives so that those high-quality images, along with snapshots from other photographers, present a vivid illustration of the hardships endured by those most affected by the panic. The most famous was Dorothea Lange's (1895–1965) photograph of Florence Owens Thompson (1903–1983) titled *Migrant Mother*. The following group of images, with original captions, offers a small sampling of this wonderful collection.

Statistics work as another effective tool by which to gain insight into the Great Depression. This chapter references five charts representing a wide range of economic indicators. With the aid of those charts, an interesting view of the Great Depression surfaces. The lengthy economic panic starts to look less like one big disaster and more like a saga. This storyline version of the

Great Depression spans seven events, from the stock market crash in 1929 to the recovery in the late 1940s. Besides giving a unique perspective of the era, the *saga* outline unravels conventional wisdom's views of the Great Depression, becoming a valuable tool for an overdue deconstruction process.

## The Unemployment Rate (Figure 2.1)

The unemployment rate ranks as one of the Great Depression's signature statistics. Prior to 1945, the Bureau of Labor Statistics' (BLS) unemployment rates were not official, with annual average estimates based on calculations by American government economist Stanley Lebergott (1918–2009) published in 1964. From January 1945 onward, the Bureau of Labor Statistics published their own official monthly figures.

Photographer Dorothea Lange's famous photograph of Migrant Mother. Nipomo, California-1936. *Courtesy, Farm Security Administration-Office of War Information.*

Figure 2.1 also shows economist Michael R. Darby's revised estimates from 1976. In the official version, the BLS treated those on work relief programs as unemployed whereas Darby counted those workers as employed. Darby's revised figures benefit the reputation of the New Deal, but even his enhanced numbers confirm the economy, as late as 1940, was still in a recession.

## The Dow Jones Industrial Average (Figure 2.2)

The Dow Jones Industrial Average measures the health of the stock market, consumer confidence and changes in the national money supply. Part of the Dow's unique value is that it has a much finer

Red cross serving beverages to men standing in line–1931. *Courtesy, Farm Security Administration-Office of War Information.*

People living in miserable poverty, Elm Grove, Oklahoma—1936. *Courtesy, Farm Security Administration-Office of War Information.*

Squatters along highway near Bakersfield, California. Penniless refugees from dust bowl. Twenty-two in family, thirty-nine evictions, now encamped near Bakersfield without shelter, without water and looking for work in the cotton. *Courtesy, Farm Security Administration-Office of War Information.*

resolution. While the other indicators post monthly, quarterly, or annual statistics, the DJIA has a value for each workday.

### Federal Government Spending (Figure 2.3)

Federal government spending dramatically climbed throughout the Great Depression, then skyrocketed during World War II, and finally settled at unprecedented peace time levels. Yet, as much as those numbers rose, they do not reflect the economic damage of an over-bearing federal government. The profusion of new regulations often created more economic havoc than any penalties imposed by high taxes and spending.

### Gross Domestic Product (Figure 2.4)

Economists designed the gross domestic product (GDP) to measure national productivity. It was supposed to be an important statistic for gauging the health of the economy. Unfortunately, the GDP fails miserably at its intended job. Instead, the indicator shadows the national money supply (included in the chart) and approximates consumer confidence. Despite its drawbacks, the GDP chart provides valuable data on the rising ratio of government spending to private-sector spending throughout the 1930s and 1940s.

### Automobile Production (Figure 2.5)

Less conventional than the previous charts, the annual volume of automobile production adds some well-needed clarity to highlight the ups and downs of the Great Depression.

## The Seven Stages of the Great Depression

The following summaries split the Great Depression timeline into seven stages. For the sake of accurate historiography, this breakdown adds lucidity compared to conventional wisdom's oversimplified Stock Market Crash-New Deal-World War II recovery model.

## Panic on Wall Street (October 1929–May 1930)

Despite Wall Street's reputation, the distressed stock market does not appear to have launched the Great Depression. Instead, the economy followed a classic, bubble-type business cycle recession. The Dow Jones Industrial Average chart (Fig 2.2) illustrates the textbook example of an oft-repeated event: the sustained rise (July 1928–August 1929), a rapid collapse (October 1929–January 1930), followed by a gradual recovery (January 1930–May 1930).

By the spring of 1930, the American economy showed some signs of a rebound. In May, the DJIA returned to its pre-bubble levels of a year earlier. The 1930 estimated unemployment rate rose to 8.7%, a significant rise, but not catastrophic. Automobile purchases plummeted, but in comparison, Detroit would not exceed the two-year 1929–1930 combined sales numbers until

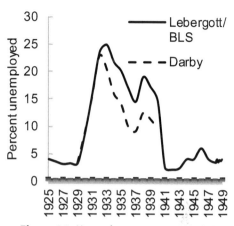

**Figure 2.1** Unemployment rate 1925-1949

**Figure 2.2** Dow Jones Industrial Average, 1927-1949

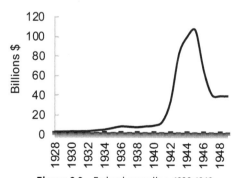

**Figure 2.3** Federal spending, 1928-1949

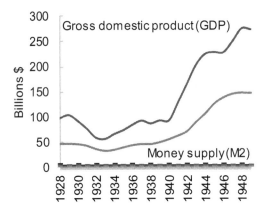

Figure 2.4  GDP-Money Supply, 1928-1949

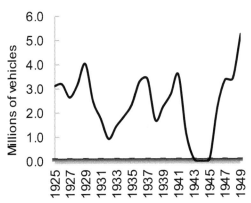

Figure 2.5  Automobile sales, 1925-1949

1948–1949. Likewise, the fall in consumer confidence caused a modest drop in aggregate prices.

Of course, the economy did not recover. Instead what should have been a short-lived recession followed by a potential *Thriving Thirties* turned to disaster as the nation's economy collapsed further into despair.

### Desperate Times (May 1930–mid 1933)

To determine the cause of the Great Depression, analysts should focus on mid-1930—not the 1920s. By May of 1930, something went terribly wrong with the economy as it plunged into a long three-year slide. At its worst, the unemployment rate topped at an estimated 25%. From peak to trough, the Dow lost almost 90% of its value. Automobile sales dropped by a staggering 75% as did the construction industry. The mainstream's standard growth metric, the GDP, fell by 40%, which approximated the nationwide average for all industries. (Note that the GDP, although unable to measure productivity, does follow short-term changes in consumer confidence.)

The period from mid-year 1930 into mid 1933 marked the heart of the Great Depression, but hard times were far from over.

### The New Deal and Beyond (early 1933–1937)

With the economy in shambles, Franklin D. Roosevelt won a landslide victory in the 1932 presidential election. Once FDR took office, he delivered sweeping changes, as promised, by implementing his highly publicized "First 100 Days" plan designed to stimulate the economy and get the unemployed back to work. The New Deal consisted of massive stimulus programs, job creation agencies, financial reforms, labor reforms, business regulations, and farm subsidies. To bolster his government initiatives and reinvigorate the

American people, Roosevelt used his strong, amiable personality, which he leveraged with the power of radio.

The economic downturn reversed, but the charts hint at the possibility that the crash had already struck bottom and that a rally had been overdue. In fact, prior to the election, during the summer and fall of 1932, the Dow rose 57% in just four months.

Whether Roosevelt energized the populace or markets naturally rebounded after they struck bottom is difficult to determine. What the charts do reveal is that the New Deal and FDR's fireside chats produced lackluster results. The unemployment rate is most indicative of the economy's struggles toward a full recovery. After a considerable four-year federal effort, the 1937 unemployment rate hovered near 15%. Automobile sales fared better, approaching 1920 levels by 1937. Considering seven lean years preceded the surge, those figures fell well short of a classic full recovery.

As per its reputation, federal spending soared during the New Deal. Like the Hoover phase, the increase in outlays, while dramatic, did not reflect the onset of new government regulations. Even though they increased FDR's political stock among his allies, the policies were harmful to many businesses.

The stock market and GDP posted significant gains, but those were distorted values inflated by the US Treasury. The Treasury had decided to add large amounts of cash to the money supply. Between 1933 and 1942, the number of American dollars doubled, and then doubled again between 1942 and 1946. In comparison, the US population from 1933 to 1946 only grew by 17%.

At least the economy looked better in 1937 compared to 1933, but then panic struck again.

## The Recession Within the Depression (mid 1937–mid 1938)

In 1937, after 4 years of a very expensive New Deal, a sudden, but short recession rocked the American economy. Historical texts on the Great Depression often omit this episode, but the sudden decline rivaled the economic catastrophes of 1929 and 1930. Between August and October, the stock market lost 40% of its value. The unemployment rate climbed from 14% to almost 20%. Automobile sales totaled 50% compared to the late 1920s.

Fortunately, the recession lasted less than a year.

## Another Recovery Attempt (1938–1941)

A sizable number of Great Depression scholars believe the pre-war recovery ended the Great Depression, but chart metrics dispute those claims. The economy tracked positive; however, much of that growth came from increased federal spending allocated for military armaments as the United States prepared for World War II. Not only did America have to rearm its own bare-bones military due to 15 years of an isolationist doctrine, but Roosevelt launched his Lend Lease policy, which provided America's European allies and the Soviet Union with valuable military hardware.

The high unemployment rate by 1940 still lingered at over 14% confirming that the economy continued to struggle. Automobile sales surged, but again, the market had more growth potential. The DJIA showed that the latest recovery attempt did not motivate investors. An increase in the GDP corresponded to the flood of military spending, not consumer spending.

### World War II (1941-1945)

President Roosevelt countered the Japanese attack on Pearl Harbor and the ongoing war in Europe with a harsh response—overwhelm the Axis powers with American military might. Through the president's directives, the nation's vast number of factories morphed into a massive military hardware and supplies machine. From small arms to giant ships, American factories pumped out prodigious amounts of war goods. The transformation won the war, but it forced the civilian economy to survive only on essentials.

From early 1942 until late 1945, much of America's non-military-related economy shut down. Besides converting factories, the war caused general commercial havoc. The axis powers controlled valuable resources and shipping lanes. Global output of consumer goods plummeted. The American government horded numerous supplies and rationed many others.

The charts on pages 19-21 illustrate the extreme contradictory nature of the war-time economy. Automobile sales highlight the dire conditions which existed in the economy during the war—zero sales for three straight years. Conversely, the unemployment rate reached record lows, and the GDP rose to spectacular heights.

Based on the stellar employment and GDP figures, another large group of academics believe World War II military spending ended the Great Depression. When combined with the pre-war consensus, most scholars place the Great Depression's final days between 1939 and 1941—but was it over?

In reality, America was at war, not in a recovery mode. For example, should the low unemployment rate achieved by millions of brave men and women serving in the armed forces define a thriving economy? The rising, skewed GDP contributed to the confusion. Most of the 275% increase occurred due to deficit spending. Meanwhile, the nation's primary end products were consumable war goods and not consumer goods.

**Recovery (late 1945–1947+)**

Post-war consumers possessed a tremendous amount of pent-up demand due to years of extreme austerity. Not only had there been recent rationing and bans on a variety of consumer goods, but over an even longer time span, since 1929, an anemic American economy had built few homes, stores, or factories. Despite the power of needing to play catch-up, a true recovery would have to overcome an assortment of obstacles.

Both businesses and workers needed to adapt to peacetime conditions. For many, it was a difficult transition. Workers, returning military personnel, and businesses had to wait while markets reshuffled.

The overreaching federal government continued to be a burden, past and present. For its war effort, the United States incurred massive debts. In 1946, federal government expenditures stood at 17.5 times greater than they were in 1929. Consequently, Americans had to compensate for both debt and spending with higher taxes. Even though the war had ended, and the New Deal was over many of their programs lingered, leaving businesses burdened with heavy regulations. Labor unions, empowered by the Wagner Act, flexed their muscles as they called for nationwide general strikes, some of the largest in American history. Labor turmoil created material shortages which plagued industries trying to move forward.

The economy's potential dire condition worried Paul Samuelson (1915–2009), the most influential American economist of the era. In 1945, Samuelson lobbied the Truman administration to launch a vast stimulus program—a second New Deal.

The statistics reflect the turmoil. The unemployment percentage rate between 1945 and 1949 rose from 2 to 6, back to 4, and then climbed up to 7.9 in October 1949, a figure approaching the mid-1930 estimated unemployment rate. The Dow surged after Germany surrendered but lost all its gains and remained flat throughout the recovery. In the confused market, a stock rally did not match GDP growth. In fact, the DJIA would not reach its September 1929 levels again until 1953. Even annual automobile sales for 1946 through 1948 continued to lag those of the 1920s.

Perhaps the historic Great Depression ended in 1941, but the recovery occurred much later. Together, the 5 line graphs suggest that the American economy did not stabilize until sometime into the late 1940s.

✧✧✧

The horrendous depth and duration of the Great Depression brought an abrupt setback to history's greatest period of economic growth. Not always a smooth ascent, the United States had faced numerous social hurdles in the past: frequent economic panics, internal and external wars, an extensive slave-labor system, race discrimination, labor strikes, and poverty. Yet, through all those obstacles the country, both economically and socially, pushed forward. Each successive generation became noticeably better off than the previous one—until the crash.

America's century-and-a-half commitment to that wonderful combination of technology and liberty produced stellar results. Escalating industrial and cultural growth created the richest populace on the planet, a society with an elevated level of egalitarianism and the greatest freedom. The economic conditions of the early 1930s altered that enlightened impression of America as the golden child, the pinnacle of civilization. From ordinary working people to academia, the Great Depression appeared to expose a crack in man's capitalist/democratic experiment, which to America's leading thinkers required at a minimum reflection, but in the extreme case, a dramatic rework.

# 3
# The Great Puzzle

Capitalism, the stock market, the Fed, the gold standard, the New Deal, World War II,—all those Great Depression topics invoke controversy. Add to the mix scholarship's myriad of diverse political viewpoints, and it is no wonder that after over 80 years of intense research, the Great Depression seems like a great puzzle. The two main political camps have their own talking points and yet they both suffer from a blind spot concerning ground-breaking industrialization.

The hub of Great Depression opinion occupies a place deep within the political left's domain, and it is those views that represent conventional wisdom. The adage "history is written by the victors" fits well with Great Depression scholarship. Progressive politicians and academics rose to power during the 1930s allowing them to reinvent government and shape the field of macroeconomics. As the decades rolled on, liberal momentum continued to dominate progressive thought in the classrooms. Consequently, generations of educators have taught their students that big government stepped in to cure capitalism's defects.

Liberal scholars suggest a variety of deficiencies in the American capitalist system as the potential causes of the Great Depression. The lead contenders include: plummeting farm prices, bank failures, and reckless speculators engaged in an unregulated stock market. Their synopsis portrays a general, pessimistic view of the US laissez-faire model. In their opinion, the Roaring Twenties received help from an artificial boost of false hope—one that the economy could no longer perpetuate. From that perspective, unfettered capitalism was overdue for a reality check—a situation in which even the slightest misfortune could send the economy into a tailspin.

Mainstream academia has a more direct explanation for the Great Depression's long duration. The New Deal started the recovery, but it took time and added stimulus to offset the damage caused by decades of an unregulated free market. Their estimate of recovery dates range between 1939 and 1941. Those who credit the New Deal as the economy's salvation lean toward 1939. The rest, and a growing majority, favor 1941 due to World War II military-industrial spending providing the jolt required to complete the recovery.

From their viewpoint, the political events in the United States that surrounded the Great Depression provide historians with a tidy scholastic cause-and-effect package. This analysis suggests that modern society needed a larger, free-spending central government in order to prosper. This is their logic. In October 1929, the month in which the economy crashed, the U.S. ran on a reduced-size federal government that included one of the leanest military forces of the century, along with a strict adherence to a gold standard. By 1946, with the war over and an economy much stronger compared to its 1920s version, the nation had done a political flip-flop. The United States federal government had upgraded to a super-sized central bureaucracy—complete with an assortment of alphabet agencies and safety-net programs. American armed forces, even after the war, remained large enough to dominate international diplomacy—the world's premier superpower. Finally, the US moved away from the gold standard, which aligned with the central bank's and treasury's new easy-money monetary policy. This viewpoint infers that a larger federal government, along with some currency flexibility, makes for a stronger and more recession-proof economy.

Expand the left's Great Depression timeline and judge it by the same criteria, and the conventional wisdom view of the 1930s unravels. The amended perspective adds the decades of the 1920s and the 1950s to the storyboard. From this alternate viewpoint, the pre-depression decade enjoyed a robust economy governed by a minimalist federal bureaucracy. Conversely, the 1930s had the indisputable combination of big government paired with a severe economic panic. For the 1940s, government growth slowed, yet the economy underwent unprecedented economic expansion. In the final decade of the timeline, the 1950s, government grew at a marginal rate as the economy continued to thrive.

Economies during the decades with smaller governments prospered, while in the 1930s, when the government was bigger, the economy struggled. That analysis aligns with economic conservative principles, an opposite conclusion than the one arrived at by the conventional wisdom crowd.

Those on the political right fault the same institutions that the left praises—big government and easy money. Conservatives blame excessive government spending and regulation as the overwhelming factors that both deepened and prolonged the economic panic.

As with the left, conservatives have uncovered an array of suspects to blame for the initial downfall. However, conservatives focus more on financial issues such as the business cycle, money supply, and the ongoing monetary fallout from World War I, which after more than a decade continued to affect the price of gold.

Conservatives blame the panic's long duration on increases in government spending, taxes, and excessive regulations. They fault Hoover for the Smoot–Hawley Tariff Act of 1930 and unprecedented peace-time spending. Right wingers consider Roosevelt's New Deal a super-sized version of all the things that Hoover did wrong.

Because some conservatives share the popular World War II recovery stance with liberals, the group leans toward a wartime or post-war recovery. Thus, the political right places the Great Depression end date between 1941 to a year or two after war's end.

Progressives command most of the scholarly thought covering the Great Depression, and they have for decades. For the most part, their monopoly has allowed them to dismiss their political counterparts as part of a *lunatic fringe*. In recent times, however, the center of political thought has made a slight shift toward the middle as conservative viewpoints now enjoy modest recognition.

The first critical mainstream treatise on the Great Depression that focused on the New Deal came from monetary expert and economic libertarian and Nobel laureate Milton Friedman (1912–2006). In 1979, PBS broadcasted *Free to Choose,* a five-part television documentary by Milton Friedman. A book by the same name became a bestseller the following year. More recent popular works include Paul Johnson's (b. 1928) *The American People* (2001) and Amity Shlaes' (b. 1960) *The Forgotten Man* (2010). All three books present an extremely negative critique of the New Deal and the people behind it. Johnson and Shlaes reached, if not a wider, at least a more contemporary audience. Those conservative pundits argued that smaller government, orchestrated by Republicans and conservative Democrats in Congress during and after the war, brought the economy back to normalcy.

Not all the viewpoints are controversial. The debate over the Great Depression shares some common ground among the experts, particularly those pertaining to the early days of the crash.

For example, the Smoot-Hawley Tariff Act of 1930, designed to lift sagging agricultural and industrial prices, attracts indignation from both ideologies. Most academics consider the extensive trade bill harmful to the economy. The exceptions are trade protectionist groups that dismiss Smoot–Hawley's destructive reputation. On the other end of the spectrum, some conservatives take the trade bill a step further claiming excessive tariffs sent the economy into a near-death spiral.

Another subject mostly independent of scholars' political affiliation is a popular thesis championed by Milton Friedman. His book, *A Monetary History of the United States* (1962), coauthored by Anna Schwartz (1915–2012), is a staple reference covering the deflationary mechanics of the Great Depression.

According to Friedman and Schwartz, the Federal Reserve Board's control of the money supply was out of phase with market forces. Their study suggests that during the latter years of the 1920s, at a time when the private sector accelerated through a boom economy, the Fed engaged in an easy money campaign instead of tightening the money supply. By late 1929 and into the 1930s, as markets fizzled, the Fed moved to tighten the money supply by raising treasury interest rates. Those final actions corresponded to a time when the economy needed lower rates.

Even those who disagree with Friedman's assessment of the Fed share a belief that the Federal Reserve Board botched their fiscal responsibility. Hence, the monetary discussion rages on. Did the Fed intervene with too little or too much stimulus, and was it too early or too late?

Pundits also agree on other fiscal issues. Many believe the failure of 5400 American banks from 1921 to 1929, and 5700 more banks between 1930 and 1932 became both a cause and effect of the Great Depression. The lingering fickleness of the gold standard also draws experts to common ground.

Political rivals continue to debate President Herbert Hoover's (1874–1964) legacy. Even though neither group supports the embattled former president, both sides concede that Hoover inherited a perfect economic storm. However, political rhetoric disagrees on how he arrived at his unflattering legacy. Those on the left rate *conservative* Hoover as a do-nothing chief executive who let the nation's economic problems fester. Conservatives blame the *progressive* Hoover for pushing an agenda that required too much spending and regulation. They consider his promotion of pro-labor legislation, farm bills, increased size of government, and sharp increases in the income tax as the original, but a smaller version of the New Deal.

Compared to Hoover, pundits view Roosevelt's presidential tenure as a more complex legacy. For his leadership role in the mammoth stimulus packages, some designed for recovery and others to promote a progressive political agenda, the left considers Franklin D. Roosevelt one of America's greatest presidents. Conversely, conservatives blame Roosevelt as the Great Depression's primary instigator. For his role as Commander-in-Chief during most of World War II (he died in office months before victory), conservatives also consider FDR one of the all-time great presidents.

ooo

Scholarship's diagnosis of the Great Depression covers a wide variety of topics, yet there are still some large gaps. For starters, how did established institutions and systems take down America's massive industrial infrastructure? Another mystery is how did a dramatic modernization occur during the recovery which followed the war? The scales of reason do not balance. On one side there are fickle investors, extra taxes, bigger government, and a mismanaged monetary system. All true. However, the other side of the equation is where pundits lose perspective. The truth is that by the time of the Great Depression American industry had developed an enormous technological footprint, one with tremendous momentum. Then, 20 years later, a post-war economy emerges that far exceeded that of the 1920s. World War II armament spending played a role, but only a minor one.

Neither side addresses the paradox. In the left's case, their fixation on big government equates to partisan nonsense. Pundits on the right have succeeded in accurately framing parts of the Great Depression problem, but their analysis lacks intensity. Where is the right's enthusiasm concerning America's twentieth-century industrial renaissance?

The Titanic-like sinking of the vibrant American economy described earlier is worth revisiting. American technology during the Roaring Twenties broke out from a jog to a sprint. Four of mankind's greatest inventions—the internal combustion engine, the electrical power grid, the field of electronics, and the motorized factory—all came to fruition. Together, they spawned a host of products that increased productivity, inspired consumer confidence, lessened the physical toil of the workplace, and thus improved America's quality of life. Sure, the economy went a bit hyperactive, and therefore a market correction to calm over-exuberant investors makes sense—but not a severe economic depression.

A business-cycle episode only explains the 1929/1930 recession. None of the standard culprits: bank failures, gold trading, excessive greed, society's faults, etc., should have had enough impact to overwhelm America's burgeoning industrial sector.

✿✿✿

In the spring of 1939, a group from New York City proudly presented a spectacular world's fair. Its attractions included an extraordinary amount of ground-breaking technology. There were man-made wonders and products for the home, farm, and factory. One of the most dramatic impacts of the fair was that those goods on display looked vastly different from their 1929 counterparts. If a fairgoer had been unaware of the Great Depression, they might assume that the American economy had flourished throughout the 1930s.

The fair did not impress or even pique most of academia's curiosity. To its mainstream critics, organizers spent too much money while the nation suffered from a financial travesty. For example, during the fair's planning stages, the unemployment rate remained at near 15%. The planners also received criticism for overestimating global harmony as World War II loomed on the horizon due to the escalation of several military conflicts in both the West and the Far East.

For their role in planning the fair, the organizers attributed the downturn in international goodwill to bad luck. Regarding the Great Depression, were exhibitors overplaying technology, passing glitter and hype off as real change? Or did the fair showcase large pockets of prosperity in an otherwise stagnant economy? A closer review of the fair validates the latter—their optimism.

# 4
# Building the World of Tomorrow

Extravagant world's fairs were popular before the age of theme parks and trade shows. Hosted by some of the world's most magnificent cities, expositions introduced fairgoers to a world vastly different from their own. Exhibits often included foreign cultures, fine art, futuristic science, the latest inventions, and renowned architecture.

Readers might be familiar with some of those fairs' iconic structures saved from demolition even though the events had shut down. London's Crystal Palace (1851) which burned to the ground in 1926, Paris's Eiffel Tower (1889), Chicago's Museum of Science and Industry, Museum of Natural History, and the Navy Pier Ferris Wheel (1893), and Seattle's Space Needle (1962) are just a few examples.

World's fairs also serve as a historical road map to mankind's rapid industrialization, as technological wizards showcased many of industry's greatest inventions. Consider a few of past fairs' most notable stars:

- New York (1853): Elisha Otis (1811–1861) debuted his elevator featuring a novel safety brake. The invention eventually made high-rise buildings practical.
- London (1862): Henry Bessemer (1813–1898) demonstrated his patented, revolutionary steel-making methods. In America, a Scottish immigrant, Andrew Carnegie (1835–1919), used the Bessemer process to launch his mighty steel empire.
- Philadelphia (1876): A massive 1,400 horsepower American-made Corliss steam engine drove an Edison direct-current (DC) generator. The electricity powered an array of incandescent light bulbs. Still in its early development stage, Thomas Edison's (1847–1931) light bulbs had enough capacity to illuminate the fairgrounds with bright lights.
- Chicago (1893): George Westinghouse (1846–1914) and Nicola Tesla (1856–1943) showed off alternating current (AC), the next generation of electric power. By winning the Chicago bid, Westinghouse won a decisive battle in the famous War of the Currents, which pitted up-and-coming AC against traditional DC, a technology backed by heavyweights Thomas Edison and J.P. Morgan (1837–1913).

- Paris (1900): Rudolf Diesel (1858–1913) introduced his more efficient version of an internal combustion engine. By the late 1930s, his namesake engine became the favored power plant for larger trucks and construction equipment. By that time, diesels also powered special-purpose locomotives.
- Chicago (1933–1934): The Union Pacific Railroad set a high-speed train record of over 100 miles per hour with its streamlined diesel-powered Zephyr M-1000 locomotive.

Despite the long history of magnificent exhibitions, one of the greatest world's fairs took place in New York City at the most unlikely time. The exposition opened in April 1939 and closed in October 1940, 10 years into the Great Depression and on the eve of America's involvement in World War II.

✧✧✧

Community leaders organized a committee to explore the feasibility of bringing a world's fair to their city during the summer of 1935. They envisioned an exposition with expansive park grounds, stunning architecture, a variety of entertainment venues, and futuristic technology. To assure the success of such a massive undertaking, the fair organizers assembled their "dream team" of men who knew how to get things done and in a big way. The all-star cast is impressive. Unfortunately, the for-profit organization lost millions of dollars—a fact highlighted by its critics. A portion of its failures came from bad luck, particularly with the timing of the war, plus it was true that the men running the fair were overly optimistic, egocentric bureaucrats. Yet, they pulled off an amazing feat. Those men were aware of an incredible array of new American technology that people who subscribe to conventional wisdom still fail to recognize today.

Committee members chose Grover Whalen (1886–1962) as president of the New York World's Fair Corporation. The former businessman and New York City police commissioner held the position as chairman of the Mayor's Committee on Receptions and Distinguished Guests. Whalen's civic duties placed him in charge of New York's famous ticker-tape parades. A notorious self-promoter, Grover Whalen went by the moniker of *Mr. New York*. In his role as Mr. New York, Whalen became a well-known diplomat.

Mayor Fiorello La Guardia (1882–1947) also served on the world's fair commission. The progressive Republican ranks as one of the best-known mayors in American history. La Guardia kept the city's economy vibrant during the Great Depression by lobbying for and receiving a hefty portion of New Deal stimulus, including funding for major public works programs. Plus, he reduced the city's crime rate by clamping down on organized mobsters.

New York City Parks Commissioner Robert Moses (1888–1981), another man with an extensive legacy, acted as the city's liaison for the world's fair committee. Today, Moses has a tarnished record as an overbearing bureaucrat who destroyed some of New York's finest neighborhoods and landmarks, including Penn Station's head house. He also gets the blame as the man responsible for sending the New York Giants and Brooklyn Dodgers to California in 1957. His high regard for the automobile and disregard for mass transportation left New York with too much motor vehicle traffic congestion and too few subway options. Yet no other person has had more of an impact on how New York City looks and functions today. His legacy includes FDR Drive, the Triborough Bridge (renamed the Robert F. Kennedy in 2013), Lincoln Tunnel, LaGuardia Airport, over 100 parks, and many community centers.

The credit goes to Moses for bringing a large corporate presence to the fair. Commercial sponsors not only introduced historic technology, but their income enabled the funding of other exhibits.

The committee employed the services of Edward Bernays (1891–1995), the father of public relations. Bernays, the nephew of psychiatrist Sigmund Freud, coined the expression *public relations* (PR) as an alternative to the term propaganda, which became more associated with wartime instead of its traditional advocate roots. At Columbia, he taught the nation's first college course on PR and wrote the first book on the subject. The field of public relations, as opposed to advertising, according to Bernays, relies on psychoanalysis and propaganda. His PR firm represented a number of popular consumer brands of the day including Procter & Gamble, General Electric, and Dodge Motors. Bernays' views embodied those held by a class of elitists who believed intelligent people should guide the *ignorant* masses. For the New York World's Fair, he served as the publicity director.

After nearly four years of planning and preparation, $160 million in spending, and the hiring of thousands of employees, the fair opened on April 30, 1939. Located at the city's geographic center in the Flushing Meadows section of Queens, the fair organizers cleared and renovated a 1,200-acre former-marsh-turned-dump. F. Scott Fitzgerald (1896–1940), in his novel *The Great Gatsby* (1925), referred to the site as "a fantastic farm where ashes grow like wheat... bounded on one side by a small foul river."

For the conversion, workers hauled 7 million cubic yards of dirt, dug the city's largest lake, and planted 10,000 trees, of which the tallest stood 50 feet high and weighed 25 tons. Along with trees, landscapers bedded over 2 million plants. They paved 62 miles of roads and paths and built 10 bridges. Workers even refurbished a harbor to accommodate pleasure crafts.

Extolling the motto, "Building the World of Tomorrow," 58 countries, 33 states, and dozens of corporations represented mankind's future. They packed their exhibits into the fair's 200 pavilions, many featuring striking architecture.

The fair operated in 1939 and 1940, from April to October. On the first day, 206,000 paid visitors and invited dignitaries toured the fair. In total, the fair attracted 45 million visitors, just shy of their goal of 50 million—much of the shortfall due to World War II.

Part carnival side show, part valuable research, infant incubators displayed at the 1939/1940 New York World's Fair. *Courtesy, Manuscript and Archives Division, New York Public Library.*

Even though the park dazzled fairgoers with fantastic thrill rides, international cuisine, lush landscaping, beautiful fountains, elegant statues, and risqué shows, along with contemporary architecture, it was the modern technologies that became the stars of the show. Vendors put on display more than just streamlined aesthetics, minimalist art, and wizardry. Visitors saw into the future—mechanical and electronic marvels that were already changing their world.

The General Motors pavilion became the fair's most popular display. GM appropriately named the scaled-down version of an ultramodern 1960 metropolis Futurama. At the time, it was the largest diorama ever built. Conceived by the famous industrial designer Norman Bel Geddes (1893–1958), the massive one-acre exhibit contained a metropolitan region. Its scale models included 500,000 buildings, 1,000,000 trees, and 50,000 motor vehicles—10,000 of them moving. Limited-access, multilane, divided highways weaved their way through a city of towering skyscrapers. The main highway artery had 14 lanes in anticipation of

General Motors - Futurama - Visitors in moving chairs viewing exhibit. *Courtesy, Manuscript and Archives Division, New York Public Library.*

looming traffic congestion. A network of smaller roads seamlessly connected the metropolis to the pedestrian-friendly suburbs that encircled the city.

GM realized that for the future, both automobiles and traffic systems demanded radical upgrades. Compared to cars built just 10 years earlier, the latest versions were faster, safer, and more reliable. With several intrastate turnpikes already operating, Geddes' vision of high-speed automotive travel was inevitable.

For the spectators, GM built endless serpentine-styled moving theater seating. The elevated conveyor seated 552 spectators who rode through the exhibit traveling 1,800 feet in 18 minutes. They sat inclined looking at the exhibit through a long window. A synchronized light and sound show accompanied them along their journey. The up-close bird's-eye view through the elaborate model simulated a low-flying airplane.

Pontiac transparent "ghost car." *Courtesy, Manuscript and Archives Division, New York Public Library.*

A second historic exhibit on display at the General Motors pavilion was a customized 1939 Pontiac Deluxe Six. For dramatic visual effect, the auto builder replaced the entire steel body with one made from a 1930s wonder material named Plexiglas. The show vehicle became known as the glass car or ghost car. The high-tech transparent shell highlighted recent automotive advances. A compact drivetrain allowed the car to have a more modern, lower stance. Nestled around the automobile's internal mechanisms, designers integrated a comfortable driver/passenger compartment and a spacious trunk. The glass panels even allowed fairgoers to see the reinforced door panels designed to protect the car's occupants in case of a side-impact collision.

These were just a few of the many changes which amounted to much more than mere luxuries. Advanced technology made the cars much more drivable. Although still an adventure, driving a 1940

RCA-Harvey Gibson, Miss Television, and James E. Robert standing with television. *Courtesy, Manuscript and Archives Division, New York Public Library.*

model automobile was less dangerous than its 1930 predecessor, thus opening the driving experience to a wider demographic.

The Radio Corporation of America (RCA) staged an impressive display of the latest communication technology. RCA exhibited the stereophonic sound quality of FM radio as opposed to the traditional static-prone AM. A facsimile machine (fax) transferred 9-inch by 12-inch documents at a rate of 18 pages per minute. But the highlight of the RCA pavilion was television. RCA placed 200 television sets throughout the fairgrounds. Thanks to the National Broadcasting Company (NBC), fairgoers viewed live television shows broadcast from NBC's Manhattan headquarters at 30 Rockefeller Center. Programs included speeches by President Roosevelt and noted physicist Albert Einstein (1879–1955). Experiments with TV technology had been going on for over 12 years. At long last, RCA's latest technology approached commercial feasibility, at least if broadcast from the Empire State Building.

At the time, most American families owned at least one radio, an invention that dated back less than 20 years. At the fair, audiences recognized the new visual media not as a slick promotional gimmick but as a high-tech product that they would soon enjoy themselves.

International Business Machines (IBM) introduced their latest electric typewriter for the growing office sector. The future computer giant also showcased the speed and precision of a punched card calculator, a machine designed to process enormous amounts of data. Even in that *pre-information age,* IBM manufactured 5 to 10 million punched cards per day.

The railroad industry, which over the previous two decades had lost its transportation dominance, brought in a splendid display of old and new technology. The Pennsylvania Railroad demonstrated the largest steam engine ever constructed—the S-1 6100. To prove its reliability and advantages, the aerodynamic bullet-nosed train ran 24 hours per day on a dynamometer (a super-sized treadmill) at an equivalent speed of 60 miles per hour.

Although not all railroad people realized it, diesel-electric locomotives would soon revolutionize railroad propulsion. Within 5 years of the fair's closing, the railroad industry built their last steam engine. The conversion from steam to diesel-electric locomotives added enough cost savings to make riding the steel rails (although mostly freight) still a viable industry. Built in 1939, the S-1 6100 retired in 1945.

DuPont's pavilion highlighted the versatility of plastics. Of the chemical company giants, DuPont was the largest, and throughout the 1930s, had been one of several developing many of the popular plastics in use today. Besides the Plexiglas used on GM's ghost car there was polyethylene, PVC, polyurethane, and many more that were already in mass production. Due to their versatility and low manufacturing costs, the Plastics Age transformed the majority of manufacturing industries.

DuPont chose the 1939 New York World's Fair to launch its revolutionary stockings made from synthetic polymers. They named the product "nylons" after the plastic's trade name. The man-made fibers proved a superior alternative to traditional silk. Women loved them. They bought an unbelievable 64 million pairs of stockings in that first year of production.

AT&T promoted the *Spirit of Electricity*. Their Bell Labs division rolled out Voder, an electronic voice synthesizer. AT&T also furnished a sophisticated television coaxial cable, which ran from the Empire State building to the RCA pavilion.

Miss Chemistry models nylon stockings at the New York World's Fair. *Courtesy, Hagley Museum and Library*

Westinghouse exhibited Elektro, a seven-foot-tall human-like robot, which under voice command could perform tasks such as walking and smoking a cigarette—the latter a popular habit at the time.

Con Edison put on a dramatic display of electrical power. In the not-too-distant past, electricity was an expensive luxury, or a novelty to be conserved. For the world's fair, New York's utility company showed how plentiful the energy source had become. Powerful generators flooded the landscaped grounds with a variety of lights such as incandescent, fluorescent, and neon. Motorized pumps at the wall of water pushed 800 gallons per minute over the man-made falls. During the summer months, electric-powered air conditioning chilled several of the pavilions.

In the Town of Tomorrow, vendors

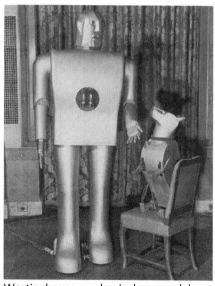

Westinghouse mechanical man and dog- -Elektro and Sparko. *Courtesy, New York Public Library*

constructed fifteen beautiful model homes. They ranged from low-cost starter houses to more expensive designs for the successful professional. Styles included both traditional and modern. Sponsors built a Fire-Safe Home from noncombustible materials. GE's Electric Home used *electric servants* to help relieve a housekeeper's drudgery. The Motor Home housed an attached two-car garage. The Plywood House had the exterior walls and roof all sheathed in Douglas fir plywood. The Pittsburgh House of Glass, as its name infers, featured generous proportions of large windows. In a nod toward the 1930s technological revolution, the glass maker noted that, "Everyone knows that in the last few years the glass industry has developed a number of products available for the builder of moderate cost homes or mansions. During that time glass products have been greatly amplified in practicability."

Theme Center - Trylon and perisphere.
*Courtes,y New York Public Library.*

For those spectators interested in *how things are made*, the fair featured numerous production lines. United Steel built a small rolling mill for the fair. Ford had a mock automobile assembly line. There was a packaging plant and an all-electric farm complete with its own food processing plant.

Fabricators erecting the fair's iconic symbols took advantage of radical welding technologies first developed just before the onset of the Great Depression. Welding allowed for flexibility and the low-cost joining of steel—a valuable technology transferable to many heavy industries. With this new technology, they built the Perisphere, a 160-foot diameter, 3000-ton steel globe supported by three columns. Liquid jets bathed the pillars in water, projecting the illusion that the large sphere balanced on a pressurized water tripod. Fair organizers named the other symbol, a 700-foot-tall steel, triangular obelisk, the Trylon.

ooo

The only remnant left on location from the 1939-1940 fair is its administrative building, which served as the original home for the United Nations (1946-1950). Later, the site hosted the 1964 New York World's Fair. Today it is named Corona Park and is home to the US Open Tennis championship.

However, the most extraordinary event held on those grounds happened in 1939 and 1940 when America's largest and the world's greatest city hosted one of history's most notable world's fairs during the nation's harshest economic catastrophe.

Today, an extensive collection of 1939 New York World's Fair archives are available to academics. These treasure troves of historical documents shed light on a couple of Great Depression puzzles. First, the world's fair demonstrated that industrialists could carry 1920s technological momentum into the next decade. Second, the fair offered a better explanation than WWII military spending as to why American technology looked so radically different after the war compared to the look its predecessors had at the start of the Great Depression.

While the Flushing Meadows exhibition helps make sense of the 1920s to 1940s industrial timeline, it adds confusion regarding the economic panic. The fact is that America somehow turned modern during those hard times. Thus, a larger mystery unfolds. With all the significant productivity gains and the many opportunities for buyer incentives, why was there a Great Depression?

# 5

# A Fresh Perspective

The technology on display at the New York World's Fair accounted for just a modest sampling of a much larger movement. Innovation soared from the 1920s through the 1930s across the entire industrial spectrum. A few scholars have recognized the revolution. Alexander J. Field, an economics history professor, and Tyler Cowen (b. 1962), an academic economist, support the modernization hypothesis. Professor Field authored *The Great Leap Forward: 1930s Depression and U.S. Economic Growth* (2012), and Professor Cowen wrote the popular pamphlet *The Great Stagnation: How America Ate the Low-Hanging Fruit of Modern History, Got Sick, and Will (Eventually) Feel Better* (2011). The pair claim that industrialization more than flourished during the 1930s; it was one of the most technologically prolific decades in American history. Yet inexplicably, the era's sudden modernization is absent from the bulk of Great Depression literature. Conventional wisdom's omission of America's innovation during 1930s casts another layer of doubt on a subject already besieged by partisan views.

To put the academic calamity in perspective, consider that two powerful socioeconomic revolutions, one industrial and the other political, coincided with America's most controversial economic disaster. The often-used adage, "there are no coincidences," suggests that there is more to the Great Depression than mainstream academia's "failed capitalism, New Deal-World War II savior" model.

A windfall of innovation combined with the New Deal's lackluster results (which included an unemployment rate that averaged over 15% from 1933 through 1939) has too many dots that mainstream academia has failed to connect. As noted in the Introduction, the story of the Great Depression needs a deconstruction—one that follows three paths: first, to track America's movement toward a modern factory economy; second, to explore how the progressive political faction addressed their concerns over industrialization; and third, to examine academia's obsession with monetary expansion while ignoring real industrial growth.

Key to understanding America's industrial journey is to cement an appreciation for our nation's political and economic dominance from its earliest days on. Alexis de Tocqueville (1805–1859), a French aristocrat and historian, noted the unique nature of the United States in his famous treatise, *Democracy in America,* published in two volumes—1831 and 1837. De Tocqueville found everything about America exceptional: its bountiful natural resources, republican form of government, economic vitality, and eclectic culture. Later, some writers particularly focused on America's meteoric industrialization development adopted the Frenchman's outlook by using the phrase *American exceptionalism*. The concept has remained relevant throughout the twentieth century and into the twenty-first century.

America's exponential growth towards modernization, although historically noted, is still underrated. In the early years of the republic, noticeable technological changes took multiple generations to occur. By the latter nineteenth century, the innovation-revolution interval dropped to a single generation and then later to a decade. With the proliferation of the electrical power grid, the internal combustion engine, electronics, and the motorized factory during the 1920s, Americans witnessed radical technological changes every few years.

By 1930, the nation's industrial expertise no longer impressed many American leaders and academics. Instead of admiration for the elegance ingrained in our capitalist structure, they saw chaos and inequality. From their perspective, in the American system, the rich get richer and the poor get poorer. In response, influential leaders and thinkers sought to slow down the pace of industrialization. They were not Luddites, but progressives did believe that unfettered capitalism caused too much economic carnage. Based on their agenda, reformists developed countermeasures to promote fairness, however because they did not grasp the benefits of America's highly mechanized economy, their policies restricted economic growth.

In those decades leading up to the Great Depression, progressives launched periodic skirmishes against the burgeoning economy. Nevertheless, the free market always prevailed. But when the stock market crashed in 1929, followed by a nationwide recession, the progressive-leaning President Herbert Hoover blamed industry for the downturn and initiated powerful reforms to halt the slide. After the economy collapsed deeper into despair, Hoover's successor, Franklin Roosevelt, doubled down with more destructive initiatives. Those also failed. Eventually, a sub-par-performing economy turned a business-cycle recession into the Great Depression.

The Great Depression ranks as the most notorious battle in a long economic war, a campaign that began not long after America's founders signed the United States Constitution, and one that continues to this day. Even though the American people had denied progressives their Utopia, which would have led to a European-centered economy, American exceptionalism did not emerge from the economic panic unscathed. Nor did the conflict end with World War

II. Most unfortunate for future economic prosperity is that the progressives finished the Great Depression satisfied that they had at least changed the rules of the game. As victors in the economic propaganda war, they gained the upper ground to write history in a way that promoted their 1930s policies.

The events that led to and then conquered the Great Depression are multi-faceted and complex. In order to understand its nature requires a study of the long-standing war between those trying to advance industrialization and those attempting to tame it. A good place to start is where American exceptionalism began—with America's booming, colonial economy.

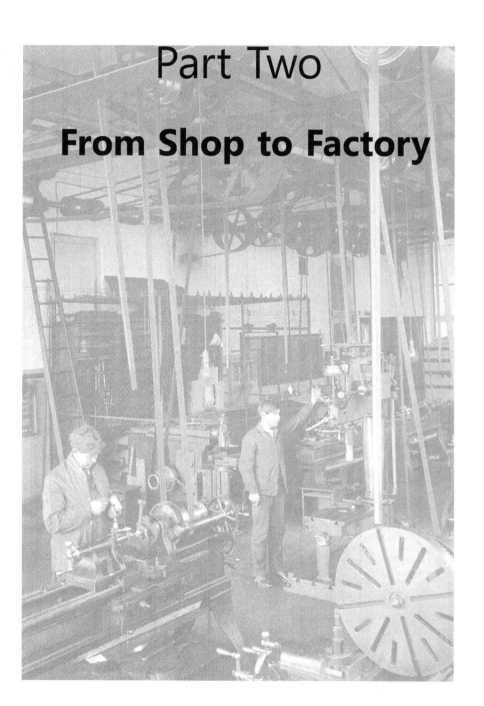

# Part Two

# From Shop to Factory

# 6

# The Robust Colonial Economy

English settlers founded the colonies with sizable advantages—a robust, diverse economy, much of which came from the continent's abundance of natural resources. To the British colonists, nature's bounty must have seemed inexhaustible compared to their homeland.

A prolific fur trade presented one of the colonists' earliest economic benefits. North America was home to an extensive variety of animals prized by the lucrative European export market for both practical and luxurious pelts. Luxury furs included mink and fox, while beavers supplied most of the staple furs. Deer, bear, and wildcat pelts were also in high demand.

However, it was agriculture that dominated the British colonial economy. Whether they were large landowners who had secured grants approaching 1,000,000 acres or small farmers with 100 acres, both groups found opportunities not available back in their homeland. Once cleared of their native dense forests, the thirteen colonies offered vast amounts of arable land with legendary depths of fertile topsoil, in stark contrast to present-day New England's rock-strewn pastures or the South's red clay—both the result of erosion from the past three centuries of deforestation.

The colonists were efficient farmers who drew upon the methods of Europe's post-medieval Agricultural Revolution. Their advanced farming included innovative iron tools, fruitful crop rotation methods, and quality livestock. The abundance of nutrient-rich land, an ideal climate for agriculture, indigenous crops developed over thousands of years by Native Americans, along with lots of farming know-how, helped farmers transform the colonies into a bountiful agricultural melting pot.

Colonial farming covered all domestic needs and generated large surpluses for export. Chief exports included tobacco, indigo (for royal blue dyes), and in the early days, rice. Cotton developed into an important cash crop toward the end of the eighteenth century.

The British partnership with the colonies further enhanced the motherland's economy. England expected to augment their national economic prosperity through the colonies in at least three areas: a supplier of raw materials; a consumer of British-made products; and a source of tax revenue. In return, the colonists received world-class protection from the British military, against France's and Spain's hostile navies.

A lean workforce was another economic bonus for the colonists. The labor surplus in England, caused by too many immigrants and failed farmers, flooded the industrial sectors with cheap labor. In contrast, America's heavily agricultural economy resulted in a labor shortage, leading to higher wages encouraging Americans to be more efficient in both their market politics and technological innovation. Nevertheless, it would be American's spirited pursuit of manufacturing technologies that elevated the colonies into a formidable society. By the 1780s, Britain's most prosperous colony had come a long way from the feudal-type society that first settled the continent from the east, especially where they were able to combine natural resources with cutting-edge technology.

With thousands of square miles of dense forests, protected harbors, rich fishing banks, and bustling international trade, Americans developed a significant shipbuilding industry. Cargo vessels were America's flagships. Although smaller than British freighters, they were ideal for trade along the East Coast from New England to the Caribbean islands.

Fast-flowing rivers, particularly in the northeast, furnished the colonies with reliable sources of mill power. Picturesque yet sophisticated, European waterwheel technology supplied efficient rotative power to a host of industries from grist mills to sawmills. For the time, American industrialization was thriving without the latest British steam engines.

In response to Europe's high demand for iron, and outside England's vision of an agrarian colony, several ironmasters built kilns in New England, with a large majority operating in Connecticut's Litchfield Hills. This area's natural resources included the three main ingredients needed for making iron: rich iron ore deposits, extensive veins of limestone (flux), and massive forests, which were ideal for charcoal. Even though outlawed under British law, colonists also manufactured limited amounts of finished iron product. Iron-making operations included slitting mills to produce iron bars and factories producing pots and pans.

A bird's-eye view provided a picture of an agrarian nation, yet beneath the region's rural exterior laid a proficient technical class. The colonists had not developed large-scale industries, yet America was a land of accomplished craftsmen. Skilled artisans included cabinetmakers, blacksmiths, coopers, printers, and cobblers, and there were hundreds more—a subculture of expertise equipped with state-of-the-art tools to assist them.

By the time of the revolution, the savvy colonists had laid much of the groundwork needed to develop a modern technical economy. In fact, the young country was already teetering on the cusp of becoming a global industrial power. Their combination of liberty, an entrepreneurial spirit, and Yankee ingenuity would soon take the nation to that next level.

# 7
# America's First Factories

The 1783 defeat of the British in the American Revolution released the aspiring country from their former manufacturing restrictions. This newfound liberty allowed enterprising businessmen the chance to pursue a new class of once-prohibited high-technology industries. Entrepreneurs placed manufacturing cotton yarn high on that list.

British-styled cotton mills offered one of the greatest opportunities to produce yarn by water-powered, labor-saving machinery. English inventor and industrialist Richard Arkwright (1732–1792) built the world's first automated cotton mills north of London during the 1770s. Cotton mills introduced western civilization to mass-produced cotton, which helped launch the Industrial Revolution. Key to cotton's success was that the end product had extraordinary potential. Cotton, once transformed from plant fibers into a manufactured textile good, became a low-cost, miracle cloth—a valuable product for the masses. A cultural game changer, cotton is a durable, comfortable, and a clean fabric with many practical uses. Its utility extended from undergarments to bedtime attire to outerwear.

America had an ample supply of raw cotton, but building a cotton-manufacturing industry

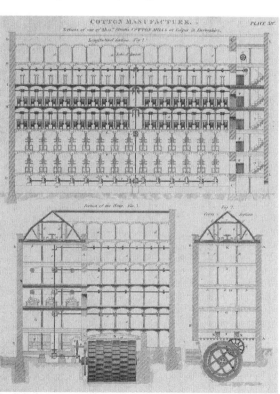

Jedidiah Strutt 18th century cotton mill.

would be a daunting task for the young nation's industrialists. Arkwright's factories had massive scale and dazzling complexity. More than just buildings, Arkwright built his factories to be machinery systems. Anchored around a network of water-powered shafts, his building stood four or five stories tall and spanned a length of over 100 feet. That ideal size became the industrial standard for the next 100-plus years. The core of his genius began with the design of an extensive network of automated machinery. It took a remarkable amount of innovation to automate the methods required to untangle, straighten, elongate, twist, and combine delicate cotton fibers. Each massive mill housed hundreds of these complex, high-speed machines.

The skills needed to build a textile mill far exceeded the technical ability of the early American artisans. Undeterred by the challenge, aspiring New England textile tycoons believed that they could produce cotton yarn as well as the Brits. After all, Americans had a strong desire for self-sufficiency, a large source of homegrown cotton, and an abundance of ingenuity.

America's first step toward mastering the art of mass-produced cotton was industrial espionage. To that aim, early textile barons smuggled in critical English textile machinery including carding machines, jennies, and water frames. They then hired recently immigrated craft workers with the technical expertise to operate and maintain that machinery.

Once installed, the pirated machines did not perform well because the European textile *experts* had only a cursory knowledge of their claimed craft. Several attempts at automated mills from Boston to New York failed, which validated the difficulty of earning a profit by mass-producing quality cotton yarn.

In 1789, undeterred by others' past failures, a company named Brown and Almy opened a water-powered cotton mill on the fast-flowing Blackstone River in Pawtucket, Rhode Island. One partner was the famous educational promoter and abolitionist Moses Brown (1738–1836). At first, Brown's mill followed a similar fate as the previous ventures—failure because of insurmountable technical problems.

Then American industry got a lucky break. As the Pawtucket mill struggled to make product, a 21-year-old Englishman named Samuel Slater (1768–1835) arrived in New York City. Slater had toiled in Manchester, England cotton mills since the age of 10, and, for the previous 7 years, he had worked as an indentured servant for cotton pioneer Jedediah Strutt (1726–1797). But Slater's was no ordinary servitude. His master worked for the Father of the Industrial Revolution himself, Richard Arkwright, as his right-hand man and minor partner. Plus, Strutt created several of his own cotton machine inventions. Young Samuel became Strutt's star pupil, more of a confident than a journeyman. In those 7 years, Slater learned the mechanical details and operating instructions of all the equipment used in an Arkwright mill. He understood the art, the science, and the math to make quality cotton yarn. Slater even became skilled in the operations and management side of running a mill. He, also, had the foresight to realize that as a textile expert, greater opportunities were available to him in America than in England.

To gain permission to leave England at the end of his service, Slater took a great risk and lied to authorities claiming to be a farmer, his father's trade. When Slater boarded a ship, and embarked on a new life in America, he left with no equipment, drawings, or even notes. He stored the mountains of cotton-making details in his head. The young man's only evidence of cotton making experience was his certificate of servitude with Mr. Strutt.

Samuel Slater arrived in New York City and soon found a job at a local cotton mill as a textile expert. The job did not last long. The poor state of the New York cotton mill was beyond repair, so he looked elsewhere for better opportunities. Slater soon learned about the ambitious undertaking at the Pawtucket mill. After some brief correspondence followed by a generous offer, the young man moved to Rhode Island and set to upgrading the Brown and Almy mill to Arkwright standards. Within the short span of one year, the mill produced its first sample of world-class cotton yarn. Slater's success launched a New England cotton mill boom. Over the next several years, textile manufacturing grew to dominate the economy of several Northeastern communities concentrated along the region's more powerful rivers.

Cotton mills established a valuable industry, but they also advanced America's industrial infrastructure and introduced a manufacturing culture to the dominantly agricultural and craft nation. This newfound manufacturing expertise helped launch America's next industrial enterprise, the mass-production of small arms.

Early America had tenuous relations with Europe's military powerhouses, England and France, which soured throughout the 1790s. Both European nations occupied controversial territories bordering the United States. For another, their powerful navies harassed American ships. Meanwhile, France was the key supplier of American military small firearms, including the army's favorite infantry weapon, the long-barreled flintlock musket.

In those early years, the young American nation maintained only a modest gun industry. Gun workshops, along with state-run weapon factories (armories) in Springfield, Massachusetts and Harpers Ferry, Virginia (now West Virginia), produced copies of English and French weapons. Those firearm *manufactories* were not true modern factories using machines and dedicated labor, but instead they housed a group of gunsmiths. In typical craft fashion, each expert hand-produced one gun at a time—from start to finish.

Because of the musket's labor-intensive design, the American army struggled to build enough arms. In response, a federal committee, headed by the US Army's inspector general, Alexander Hamilton (~1756–1804), sought a private sector partnership in a bold plan to produce muskets in higher volumes and at lower costs. The aggressive strategy required mass production methods. A musket's combination of complexity and precision made it a worthy candidate for high-volume manufacturing.

There were other factors favoring machine manufacturing. A musketeer did not demand a personalized artistic flair (a must-have by gun-toting aristocrats). Muskets were less complicated than a rifle, but the larger volumes required by the military qualified them as an excellent choice for mass production.

For ease of field repairs, its sponsors demanded a musket's components to be interchangeable. This meant that every feature of those selected components had to meet tolerances specified on a drawing. In theory, the plan eliminated the tedious tasks of hand fitting to match mating parts. A musket contained approximately 50 parts, known as "all lock, stock, and barrel." For interchangeability to work, each component had to be a near-match to one made by another worker, on another day, using a distinct set of tools and gauges.

In 1798, Hamilton's group awarded the army contract to Eli Whitney (1765–1825), the already-famous American cotton gin inventor. The order demanded the delivery of 10,000 muskets within four years. Whitney's labor-saving cotton gin had created a supply boom of raw cotton at a critical time, but the oft-copied gin did not produce much income, so Whitney abandoned the cotton industry and instead turned to arms manufacturing. The great inventor entered the agreement with minimal skills as a gunsmith, but what Whitney lacked in knowledge he made up with forward-thinking and a can-do attitude.

Historians know little about Whitney's operation except that for its time it was impressive. The most detailed information describing the gun maker's factory complex comes from a letter written by Whitney's ten-year-old nephew, Philos Blake (1791–1871). The young man wrote, "There is a drilling machine to bore barrels and a screw machine...a blacksmith shop and a trip hammer, and five hundred guns done."

Eli Whitney earned a reputation as a great inventor, but he also developed a knack of capitalizing emerging technologies, both foreign and domestic. French and English armories had pioneered cutting-edge machinery, jigs, and gauges. In his own back yard, New England mills employed waterwheel technology, which included a system of shafts, belts, and pulleys capable of transferring mechanical power to the machinery. Whitney pulled able machine builders and machinists from cotton industries. He packed his network of labor-saving machines into a factory complex located outside New Haven, Connecticut.

His goal was to use semi-skilled operators and division of labor principles to replace the limited number of available higher-skilled artisans.

As a visionary, Eli Whitney might have been too far ahead of his time. To complete the immediate task required overcoming numerous technical obstacles. Because he underestimated the enormity of the challenge, Whitney's plant took 10 years to fulfill the original four-year contract, while the necessity for a few hand-fitting operations meant that some components were not interchangeable. Still, Whitney earned a second station in American history as he helped set a strategic path toward mass manufacturing.

By the 1830s, with contributions by such gun makers as Simeon North (1765–1852) and John Hancock Hall (1781–1841), the young industry had mastered small arms standardization with interchangeable parts, which elevated America's gun factories to the most advanced in the world. Historians refer to those methods of industrial proficiency as *The American System of Manufacturing*.

In 1789, the United States not only introduced a revolutionary political system, but as in Europe, the transition from farm to a factory radicalized society. For centuries men had worked the land, or they took on a vocation as a trader, blacksmith, carpenter, or banker, but rarely did they become factory workers. Nor did women labor in commerce. Instead, they raised children and made, repaired, or grew what the family could not afford. They also managed the home. The pre-industrial housewife lived a hard, hectic, and demanding life. So, even as cultural changes caused conflict, the benefits of industrialization, which included greater health, comfort, and liberty, became a worthwhile pursuit.

# 8

# The American Industrial Revolution

To mass-produce musket components at interchangeable standards, American gunsmiths advanced a family of industrial machinery known as machine tools. Machine tools cut or form materials, usually metals, to precise measurements. At the time, machine tools were in their infancy. English ironmaster John Wilkinson (1728–1808) built the first documented machine tool just a generation earlier. Wilkinson's boring machine, built in 1774, cut a steam engine's large cylinder bore. Soon, other emerging industries developed machine tools to advance their products. Led by the British and French, their mechanics invented simple versions of several common metal-removal machine tools. Those late eighteenth to early nineteenth-century machines included the drill press to machine holes, the lathe to process cylindrical parts in a rotating chuck, and the milling machine that makes cuts on non-cylindrical work held firmly in a vise.

Americans also pursued the design and construction of European-styled machine tools. Despite Europe's formidable head start, America's early class of machinists, led by gun manufacturers, played a quick game of catch-up with their European counterparts. By the mid-nineteenth century, Americans had built the most prolific machine tool industry in the world. Theories vary on how a young nation gained superiority over the masters. Candidates included America's labor shortage, a can-do attitude, Europe's penchant for using handcraftsmanship over automation, or most likely, all the above.

American machine tool builders embraced a passion for continuous improvement. Iron bases replaced wooden frames for greater robustness. Machines became larger and more versatile. With the aid of basic hand tools, ingenious technicians developed methods to improve the accuracy of precision mechanisms. These, and other upgrades, elevated their machine tools into high-performance, elegant products that were sturdy, adaptable, and precise.

In both the fields of metal removal and metal forming, machine tool manufacturers also expanded their product lines. They introduced new types of equipment to cut metal, such as planers and shapers, and metal forming machines to forge, press, stamp, draw, swage, and roll metals.

Until the mid-nineteenth century, store-bought machine tools were not available. Instead, factories often used farmers-turned-mechanics to build machines in-house. Eventually, the market for machine tools grew large enough to support commercial companies, and machines became more general purpose. Just before the Civil War, the list of successful machine tool companies included A.M. Freeland (New York), Pratt & Whitney (Connecticut), Brown & Sharpe (Rhode Island), Warner & Whitney (New Hampshire), and Robbins & Lawrence (Vermont).

America's technical and commercial advancement of machine tools produced dramatic economic results. Manufacturers refer to their machine tools as *machines that make machines*. Thus, the widespread availability of machine tools gave innovators the opportunity to prototype their inventions and then mass-produce them. The greater accessibility and utility of general-purpose machine tools made the assembly line with interchangeable parts practical for a wide range of industries. From that era, besides small arms and textile machinery, came the:

- Telegraph, 1837 – Samuel Morse (1792-1872), New Jersey
- Reaper, 1845 – Cyrus McCormick (1809-1884), Chicago
- Sewing machine, 1846 – Elias Howe (1819-1867), Massachusetts
- Rotary printing press, 1847 – Richard Hoe (1812–1886), New York City
- Rotary valve stationary steam engine, 1849 – George Henry Corliss (1817–1888), Rhode Island
- Safety elevator, 1852 – Elisha Otis, Connecticut

Thanks to England's Industrial Revolution, America gained expertise in cotton mills, armories, agriculture, machine tools, and other industries; yet, Great Britain gave one more significant gift to the industrialized world—the steam engine. Engineering genius, James Watt (1736–1819), and his financial partner, Matthew Boulton (1728–1809,) introduced their novel steam engine in 1776. A tremendous leap in engineering, scholars cannot overemphasize the Bolton and Watt engine's historical impact. It was the size of a house, as complex as a spinning jenny, yet built with the precision of a Swiss watch. At the time, nothing matched its sophistication. The steam engine inspired the first machine tool (the Wilkinson boring machine mentioned earlier in the chapter), the field of mechanical engineering, and the trade of the toolmaker. When a reporter asked Matthew Boulton about the nature of the enormous contraption under construction at his Birmingham foundry, he responded, "I sell here sir what all the world desires to have—power."

Watt designed his steam engine to operate water pumps installed in flooded coal mines. The steam engine later found use as a factory power supply where mighty flowing water was unavailable or unreliable. Other resourceful inventors used the versatile steam engine to power ships and trains.

In 1807, American artist and inventor, Robert Fulton (1765–1815), applied a Boulton and Watt 20 horsepower, double-acting, rotative steam engine to power America's first commercial steamboat. Fulton's *Clermont* ferried passengers up and down the Hudson River between New York City and Albany. Within two decades, American shippers provided freight and passenger service over dozens of lakes, rivers, harbors, and coastal waterways. By the 1850s, steam propulsion freighters with sail assist gained an economic advantage over pure sailing ships on transatlantic voyages and California Gold Rush-inspired transcontinental cruises (interrupted by freight and passenger transfer over the Panama isthmus). This steamboat era produced one of America's first mega-wealthy tycoons, Cornelius "Commodore" Vanderbilt (1794–1877).

Along with steam power, early nineteenth-century America had a developing system of state-built canals constructed to augment the nation's many slow-moving waterways. The most famous was New York's Erie Canal (1825). Canals used basic but effective methods to move up to 30 tons on a single barge. Despite their capacity to haul large loads, railroads quickly doomed the canals' usefulness.

The rise of railroads is one of mankind's major triumphs, a merging of numerous technologies that reinvented themselves for almost 200 years. Railroads grew in popularity throughout the late eighteenth and early nineteenth centuries in both England and the United States. In the days before steam engines, railroads used either horses or oxen to provide the pulling force. By the early 1800s, James Watt's steam engine inspired a few adventurous entrepreneurs to pursue steam power's land-transportation potential. The original steam railroads employed a system of inclined planes. A stationary steam engine, attached to an endless rope, pulled the rail cars up a grade, while on the other side gravity propelled the payload down the track.

In 1825, British inventor George Stephenson (1781–1848) took the great leap and completed the first effective test of a steam-powered locomotive. Inspired by Stephenson's triumph, a few enterprising Americans tested versions of steam trains for passenger service. In the summer of 1830, Peter Cooper (1791–1883) had modest success with his small, experimental Tom Thumb locomotive, which he tested on 13 miles of track for the Baltimore & Ohio Railroad (B&O). Later that year, on Christmas Day 1830, a locomotive named the Best Friend of Charleston steamed out of Charleston, South Carolina. The wood-burning engine rolled over six miles of wooden rails capped with iron straps, establishing America's first commercial rail line. A few days later, a local newspaper reported on the historic excursion:

> The one hundred and forty-one persons flew on the wings of wind at the speed of fifteen to twenty-five miles per hour, annihilating time and space... leaving all the world behind.

Progress would soon take railroads to another level. Engines became more powerful as coal replaced firewood. The upgrade to solid iron rails supported heavier loads. Railroads provided great speed and comfort for people, but they also became practical for moving freight.

The steam engines' imposing presence, which featured a distinctive style of form following function, helped define the vast American landscape. The puffs of trailing smoke, the whistle of a locomotive's steam horn, the clanking of wheels rolling down the tracks, the single bright headlight, and even the cowcatcher—all those sights and sounds became iconic Americana.

They were also dangerous, loud, noxious, and sometimes unreliable, but by 1850, powerful steam locomotives traveled on over 9,000 miles of track throughout the Eastern seaboard states. Over the next 10 years (by 1860) the length of railroad lines more than tripled to 30,000 miles of track. This revolutionary industry by then had set its sights on a transcontinental road.

The advancement of mid-nineteenth century technology pushed the American economy to the verge of something big. Oceangoing steamships supplied important foreign raw materials and tapped a valuable export market. Railroads transported great volumes of raw materials into factories and mills and large quantities of finished goods out. Commercial machine tools made large-scale mass-production feasible across a variety of industries. This combination of the machine tool industry with a vast network of rail and shipping lines provided critical prerequisites for a true national, industrial strategy.

That combination of mechanization and railroads unleashed a capitalist rush never before achieved. New technologies, paired with producers, retailers, and consumers, formed a massive commercial network. A sewing machine manufacturer in Connecticut had access to every major population center across the country. Strawberry growers in Virginia could expand their operations knowing they could deliver fresh berries to Chicago, New York, and dozens of other cities. Ocean-going steam vessels brought the commercial expansion to a global scale.

Historiographers declared the American technological movement the beginnings of a new era, often referred to as the Second (or the American)

Industrial Revolution, as it far exceeded England's dramatic factory transformation 75 years earlier.

ööö

Despite the nation's new enlightened status as a wealthy global power, Americans still needed to address a cultural tragedy—the brutal institution of slavery. Even though America's abolitionist crusade made significant progress, cultural differences heightened the tensions between anti-slavery and pro-slavery supporters. In 1860, the rise of abolitionists had matched the intensity of those advocating slavery. The anti-slavery movement may have been on the verge of overwhelming pro-slavery attitudes, but social and political factions failed to negotiate a peaceful solution. In 1861, the Civil War divided the nation.

A remarkable correlation exists between the rise of industrialization and the decline of slavery. The lead hypothesis credits mechanization's trend toward high productivity and a need for specialized trades, which minimized the benefits of slave labor. Alternatively, a population that is comfortable and secure might indicate a social element to the problem—society's growing compassion for their fellow man. Speculation aside, the timing is telling.

Some of society's most cherished written works, including classic philosophy, religion, or the renaissance do not condemn slavery, at least outside of their own kind. There were anti-slavery laws and proclamations introduced from the thirteenth through to the eighteenth centuries, but many were conditional or overturned. Western Europe's industrial revolution, beginning in the 1780s, marked the first dramatic shift in mankind's perception of slavery and cemented a growing outrage against enslavement coupled with a commitment to the abolitionist movement.

In a young United States, the emerging factory economy paralleled anti-slavery views. Figures 8.1 and 8.2 track changes in the slave populations of four regions. For the northern states (Figure 8.1), a sharp decline in slavery began after 1790. That was the year the US Constitution took effect and the age that the North industrialized. That year, the northern states held over 40,000 slaves. As the abolitionist movement gained momentum, those states strengthened laws that outlawed slavery. Slavery mostly disappeared from the region about midcentury. The lone holdout was New Jersey, with the 1850 census counting 236 slaves.

The eastern border states, positioned between the North and the South, absorbed some transference of northern attitudes (see Figure 8.1). With those states, a significant decline occurred after 1810 as manumission grew more acceptable.

Much less industrialized than the North, the upper southern states moderated the growth of their slave population, yet they maintained their slave culture—see Figure 8.2. While industrialization discouraged slavery in the north, it encouraged the practice in the deep South. The North's cotton production boom increased the demand for raw cotton. As Figure 8.2 illustrates, the states with the least industry grew the most cotton causing a surge in the need for more slaves.

While increased economic activity maintained a high demand for slave labor, the machine tended to civilize mankind. As the nation's 1850s factory economy intensified, anti-slavery viewpoints became intolerable by the majority. With or without war, the United States would end slavery.

The relationship between the rise of abolition and industrial progress would continue to repeat itself numerous times around the globe. In time, and at a much slower pace, modernization would also lessen racial hatred and prejudices.

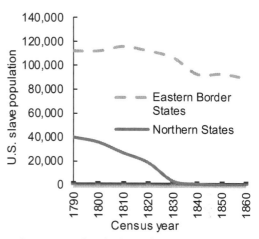

**Figure 8.1** The decline of slavery in the North

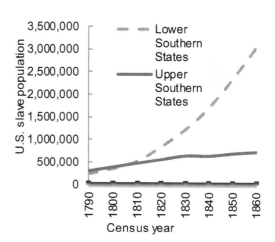

**Figure 8.2** The rise of slavery in the South

# 9
# The Rise of Big Business

Within a decade or two after the Civil War, the Industrial Revolution ended its reign. Machines no longer appeared rebellious. Citizens just accepted the steady introduction of technological marvels as routine. Mark Twain (1835–1910) sarcastically labeled this follow-up industrial economic era the Gilded Age. The great writer and humorist penned the catchy, popular term as a moniker for the late-nineteenth-century era noted for its financial generosity to an elite group of industrialists, railroad men, and investors.

This era can claim some of America's most famous tycoons, with the core of the class born in the 1830s. Rail titans included Cornelius Vanderbilt, who first made a fortune in shipping, and Jay Gould (1836–1892). The financier J.P. Morgan (1837–1913) also dabbled in rail before he got into steel and later the national power grid. A Scottish immigrant, Andrew Carnegie built the world's first high capacity steel mill, making him wealthier than any of the railroad tycoons.

Metalworkers for hundreds of years knew steel as a wonder material because of its superior properties of strength, hardness, and durability, yet practical mass-production and lower costs proved elusive. On a trip to England, Carnegie discovered an innovative steel-making furnace invented by Englishman, Henry Bessemer. Based around the Bessemer process, the Scotsman launched a steel-making empire anchored in Pittsburgh, Pennsylvania. Builders first adopted Carnegie steel to replace iron used for train rails and bridges. Architects erected the first skyscrapers based on steel's superb strength and versatility. Later steel became a staple for machinery, automobiles, and appliances.

Right on the heels of America's first oil boom, John D. Rockefeller (1839–1937) and his partners built the Standard Oil Company. Standard Oil refined and sold quality, lantern-grade kerosene at affordable prices. Due to its large size, research funding, and high efficiency, the oil giant took the most credit for lighting up American homes, businesses, and urban areas. As a result, Standard Oil's petroleum made a major impact toward improving the quality of life for millions of people when there was little or no sunshine available. Thanks to the success of his refinery operations Rockefeller became the wealthiest of all his fellow Gilded Age titans. The Supreme Court broke up Standard Oil in 1911. The oil giant had been so efficient that for decades it had limited competition.

There were many more influential entrepreneurs creating hundreds of significant enterprises, which allowed America's industrial base to thrive. The Gilded Age bridged the American Industrial Revolution to the modern, hyper-factory age of the twentieth century. Firms of that era were large but not to the scale of present-day corporate giants. For example, instead of hired guns, men like Rockefeller, Carnegie, and near the end of the era, Henry Ford (1863–1947), founded and ran their companies. They manufactured their goods in-house, a system known as vertical integration. For some, this strategy grew out of necessity as cutting-edge technologies pushed the limits of the nineteenth-century industrial infrastructure. It took decades, but a vast vendor network evolved that could supply commodities and specialty items to the manufacturing community. Today, an outsourcing strategy, or horizontal integration, allows big corporations that are too cumbersome for vertical integration to prosper.

Mass retailers emerged during the late nineteenth century serving as both an indicator and an enabler of the period's economic progress. With more affordable goods available and ample consumer confidence, retail chains replaced traveling merchants and general stores. Department stores such as Macy's (1858) and Gimbals (1887) anchored their flagship stores in big cities. These urban giants sold products that spanned from bargain to luxury brands. Smaller cities and larger towns attracted five-and-dime retailers such as F. W. Woolworth Company (1878) and later W. T. Grant (1906). Those stores also sold a variety of wares. The mail order retailers, with the help of the United States Postal Service, had a tremendous influence on a still mostly rural America. Montgomery Ward launched the concept in 1872 while Sears & Roebuck (1893) expanded the model, becoming the most influential retailer in its market. The extensive Sears catalog offered everything for both home and farm.

Retail stores served a larger role than just factory sales agents. Chains, especially Sears, often challenged manufacturers to develop a better product at a lower price. In return, the retailers assured their manufacturers that higher volumes would earn the extra revenue needed to cover their investments and still allow them to make a nice profit.

A near-perfect economic environment spawned Gilded Age prosperity, which allowed American ingenuity to produce a constant stream of innovative technology. Improved financial systems pooled resources for the all-important investor. A hands-off-the-private-sector structured government respected an individual's liberty. A vast and educated American labor force specialized in a variety of skilled trades. Much of that technology crossed sectors, creating a massive cutting-edge network sharing knowledge across many industries.

Contrary to its reputation, the Gilded Age was beneficial for ordinary Americans. Gilded Age tycoons did not pilfer a large slice of the economic pie; instead, through innovation and expansion, they enlarged the pie. Factories made quality products that sold for lower prices. The modern economy supplied steady work with good wages for millions of workers. As a result, the era enriched Americans quality of life.

The Gilded Age industrial chiefs also left behind an impressive legacy of philanthropy. Rockefeller and Carnegie each gave away a large part of their fortunes to charity, a sum that in today's dollars rivals the net worth of the world's wealthiest multi-billionaires. These men funded programs that helped the poor and ended diseases. Lots of their benevolent investments in education and culture produced long-term benefits. For example, Andrew Carnegie funded the construction and stocking of 2,509 libraries. Despite the volumes of documentation chronicling Rockefeller's benevolence, researchers continue to uncover more of his wonderful deeds. Today, a traveler visiting a major museum, a national park, or a historic landmark, probably are benefiting from good deeds by Rockefeller or other Gilded Age tycoons.

As every pupil learns in school, historians and social critics have not been kind to the titans. Gilded Age critics launched a widespread crusade against industry. Contemporary journalists, notably the muckrakers, became obsessed with other people's wealth and power. The media assumed that tycoons did not earn their fortunes. Serial magazine articles and books chronicled what they perceived as unethical business practices. Newspapers published personal fortunes to shame wealthy industrialists. Progressive politicians passed legislation designed to punish companies for being large and successful. The nation's judiciary, including the Supreme Court, considered the productive act of lowering prices through innovation as restraint-of-trade violations—a serious crime. Then, as they have done for generations since, mainstream academics continue to perpetuate the derogatory term for the era's tycoons as *robber barons*.

Along their path to modernization, the American people encountered obstacles. So, it is no surprise that 125 years ago, a still young, capitalist, work-in-progress America faced its fair share of heartfelt social issues. Yet the true historical lesson we learn from the Gilded Age is not about the gross excesses of the rich or the struggles of the impoverished, but how it cemented the axiom that America continued as the land of wealth, liberty, and opportunity.

By the turn of the century, the Gilded Age approached the end of its reign, only to be usurped by an era that produced new technologies at an even faster pace. The prolific twentieth century introduced the electrical power grid, wireless communication, automobiles, airplanes, and other game-changing technologies.

# Part Three

# A New Century with New Technologies

# 10

# Launching Innovation

Nineteenth-century technologies, although impressive, served an even greater purpose by launching a new collection of modern marvels for the next century. By the 1920s, capitalism's nth-degree economic growth rate had reached an unprecedented pace. The star of the diverse group of technologies was the polyphase electric grid. This energy source revolutionized cities, rural areas, homes, businesses, and farms. However, there were many other stars. Technology revolutionized transportation. Over highways, on rails, in the air, off road, and across the seas, mechanization intensified mankind's ability to move people and goods. Pioneering electronic gadgets included radios and experimental television.

For the homeowner, labor-saving appliances such as refrigerators and washing machines reduced the workloads for millions. Food processing plants produced consumer goods that were both nutritional and convenient. Material mills increased their output of specialty steels, stainless steels, and aluminum alloys, along with some rudimentary polymers. Each material had unique advantages that allowed manufacturers to supply a vast array of products with greater utility, performance, and affordability.

The introduction, improvement, and expansion of thousands of engineering marvels turned the 1920s into a futuristic world that transformed American society. Those new technologies improved the quality of life for the bulk of Americans. British historian Paul Johnson summed up American technology in the years prior to the Great Depression this way: "...blessings denied in adequate quantities...to most of humanity for countless generations...were suddenly made available in less than a lifetime..." Yet despite all of industry's accomplishments, somehow the Great Depression followed the Roaring Twenties.

The prevailing schools of thought blaming an over-valued economy for the Great Depression are at odds with the 1920s hyper-capitalist culture which fueled industrial advances beyond anything in history. True, the timeline fits. The long recession followed the Roaring Twenties, but so did high tariffs and the New Deal. What the 1920s industrial model provides insight into is the recovery mystery. The ongoing high-tech, high-production 1920s economy rationalizes the dramatic modernization of the late 1930s and the 1940s.

From one social science to another, macro-economists hampered historians' attempts to understand the Great Depression. Economic wealth and growth are abstract concepts, but economists conceived a method to measure these values—the gross domestic product (GDP). The Keynesian-inspired GDP depends on two factors—the business cycle and monetary expansion. Neither factor accounts for the rapid advances in economic wellbeing provided by technology.

The GDP fiasco gets a detailed explanation in later chapters, but for now realize that historians have lacked the right tool for the job. To add clarity to the discovery process of the wealth puzzle, the following chapters examine a few examples of the many technologies responsible for the launch of the most prosperous decade in world history at that point. The model of the 1920s discussed in these chapters will reveal a vibrant pro-capitalist economy as opposed to the lackluster hypotheses cited in popular historical accounts.

So, how robust or *not so distressed* was the pre-Great Depression economy?

# 11
# Henry Ford's Model T

The proliferation of the automobile, which by the mid-1920s surpassed 20 million vehicles produced, dramatically elevated Americans' mobility. The remarkable utilitarian value of the automobile as an effective, personal method of transportation ranged from essential needs to recreation. Autos allowed more people to visit the vast wonders of a diversified nation. They encouraged the population to spread out, enabling an explosive migration to the suburbs—a compromise between city and country living that seemed to fit the needs of the modern American family. Motor vehicles brought producers and consumers closer together through a growing number of retail stores and farmers' markets—on their own a major boost to American economic output. Finally, the popularity and complexities of automobiles initiated a wave of technological innovation that crossed over to other industries, raising the proficiency of all manufacturers.

The automobile's major obstacle was its preferred power source, the internal combustion engine (ICE). Advanced engine design development occurred over a period of decades and involved many individuals, most of them unknown. One notable inventor, a German, Nikolaus Otto (1832–1891), boosted its thermodynamic efficiency with his invention of the four-stroke cycle, which included a compression and a power stroke for greater efficiency. An arrangement of pistons, a crankshaft, and valve linkages, driven by a camshaft, optimized the Otto cycle. Some other breakthrough engine innovations included: the carburetor for fuel delivery, a high-voltage circuit for ignition, and the ideal fuel, an inexpensive, readily available petroleum byproduct named gasoline. Engines used fluid bearings for many of their rotating parts. Fluid bearings are space saving, inexpensive, reliable, and they reduce vibrations.

Both the automobile and the internal combustion engine came to America through France and Germany. Due to its immense complexity, the ICE took the longer journey from prototype to mass production. When the first engines reached the United States in the 1890s, the automobile's *power plant* still required further development to become a useful technology. Early competition from electric and steam-powered vehicles attests to the market's uncertainty on the future of automobile propulsion. But the gas engine had enormous potential. It required but a few years for a pack of mechanics to solve the inefficiency problems and eliminate steam and electricity as automotive power-supply contenders.

Despite its folksy moniker, 'the horseless carriage', autos borrowed much of their technology from the bicycle, at the time a popular means of transportation and recreation. By the turn of the century, the bike embodied a broad spectrum of the latest in mechanical engineering. The rigid, lightweight frame contained high-tech alloy-steel tubing brazed together into a strong configuration. Wheels, with laced and tensioned precision-made spokes, supported crafted wooden rims fitted with pneumatic tires for a smooth ride. A clever, efficient chain-and-sprocket drive delivered power from the cyclist to the rear wheel. The wheels, crank, pedals, and steering tube all smoothly rotated on the recently invented ball bearing, which helped make the total package extremely efficient. For the transition from bicycles to automobiles, builders added two more wheels, a bench seat or two, installed an engine coupled to a beefed-up bicycle drivetrain, and America had its first automobile.

Whereas the Europeans introduced automotive technology to the world, America harnessed its potential. One man earned the bulk of that early credit, a mechanical genius from Michigan and one of the most recognizable names in American industrial history, Henry Ford.

Henry Ford grew up on a farm outside Dearborn, Michigan, twelve miles from Detroit. Rejecting the family's farming tradition, an 18 year-old Henry set out for adventure in industry. Ford spent the next 24 years educating and preparing himself for his future role as the man who would produce the "perfect" automobile. As a young man, the versatile Henry Ford learned a trade as a machinist. Later, he completed bookkeeping courses. The first part of his training taught him how to build cars, and the second, how to run a business. He eventually became technically proficient enough to build his own operational internal combustion engine—an impressive accomplishment. Not just well educated and driven, he must have also been daring. Henry Ford raced automobiles, in part as a publicity stunt designed to promote his name recognition, but also to gain greater insight into state-of-the-art automotive designs.

For much of the 1890s, a thirty-something Henry Ford worked as an engineer for the Edison Illuminating Company in Detroit. There he caught the attention and support of the master inventor himself, Thomas Edison (1847-1931), who encouraged him to pursue and develop automotive technology. After several years of spare-time, hobby-car building, Henry Ford, in 1899, turned entrepreneur. That year, Ford and his backers formed the Detroit Automobile Company, which failed before it could get off the ground. With his reputation intact, Ford with a new set of partners established the Henry Ford Company in 1901, but that partnership did not work. Ford resigned within four months (the Henry Ford Company morphed into the Cadillac Automobile Company). Then in 1903, two weeks shy of his fortieth birthday, Henry Ford, along with his investors, including the Dodge brothers, established the Ford Motor Company.

1913 Ford Model T touring car. *Courtesy, The National Museum of American History, Smithsonian Institution.*

Initially, Ford built his cars in a similar fashion to the hundreds of other automotive manufacturing firms of the era; he hand-built each model one vehicle at a time. However, Henry Ford differed on his marketing approach. Ford's major competitors considered the automobile a luxury item—a product affordable only by the well-to-do. In contrast, Ford decided early on to pursue a low-cost vehicle. The strategy to make an inexpensive, working man's car succeeded. In its first five years, the Ford Motor Company sold nearly 20,000 vehicles. Its evolving product line, models A, B, C, F, K, N, R, and S, made Ford one of the largest car manufacturers in the country, but Henry Ford saw the potential to sell a much greater number of automobiles. As the famous technologist explained,

> I will build a car for the great multitude. It will be large enough for the family, but small enough for the individual to operate and care for. It will be constructed of the best materials, by the best men to be hired, after the simplest designs that modern engineering can devise. But it will be low in price that no man making a good salary will be unable to own one-and enjoy with his family the blessing of hours of pleasure in God's great open spaces.

Following that vision, the driven industrialist redirected the company to focus on just one model. In October 1908, the five-year-old company launched the *universal* automobile, the iconic Model T, a low-priced, reliable car built for the average consumer.

Ford's Tin Lizzy, the Model T's nickname, met all the auto tycoon's rigorous criteria.

The Model T's performance specifications included a four-cylinder, 20-horsepower engine capable of 40 miles-per-hour and a fuel economy approaching 25 miles per gallon. Designed for rural America, a Model T had ample ground clearance and nimble suspension, an ideal combination for negotiating deep-rutted roads, climbing steep grades, and crossing shallow streams. For comfort, some Model Ts came equipped with an enclosed cab. The new Ford even featured electric lights for nighttime driving.

Most impressive about his vision is that to Ford, low cost did not mean cheap parts or shoddy production. Henry Ford's Model T was tough. The innovative automaker made the chassis, along with other high-stress components, from vanadium steel. At the time, only race cars used the high-strength, exotic alloy. Durable steels like vanadium were difficult to manufacture, especially in high volumes. The more Ford's engineers gained expertise working with specialty steels, the more applications Ford found for them. Vanadium steel cost more, and due to its toughness, and despite a growing expertise to process it, was more expensive to turn into automobile parts. For Ford, the benefit outweighed the extra cost as vanadium steel made the car stronger and lighter—an overall higher performer.

Besides steels Henry Ford used top-of-the-line tires, glass, leather, and electrical components. The generous use of high-quality materials is one reason so many Model Ts stayed on the road for so long.

A well-known nuisance associated with the Model T is that it required a hand crank to start the engine instead of an electric motor. To Henry Ford, the crank starter was a positive feature—he believed that even the best batteries of the day would have been a weak link in his otherwise robust car. It was important to Ford that the Model T maintain its reputation as reliable while being inexpensive to operate. Ford offered an electric starter motor beginning in 1919 but only as an option.

An instant success, Ford sold 10,660 Model Ts in its first model year (1909). Three years later, as high production techniques kicked in, Ford sold nearly 70,000 of the 1912 model vehicles. The Tin Lizzy's success prompted Ford engineers to adapt creative processes to meet the increasing demand.

As sales for their automobiles swelled, Ford engineers, in part driven by their boss's passion for continuous improvement, upped their use of automation to increase output and reduce unit costs, resulting in lower prices, which, in turn, generated greater sales.

One of those revolutionary production methods led to the moving assembly line, a manufacturing system in which workers performed their tasks along a moving conveyor. For the Ford Motor Company, the change from division-of-labor methods to a moving assembly line required several incremental steps over a six-year period. In 1908, one of Ford's top engineers, Charles "Cast-Iron"

Sorenson (1881–1968) (Sorenson acquired his nickname from his earlier days as a maker of sand patterns to mold cast iron), developed an assembly line to speed up construction of the labor-intensive magneto used on the Model N. A magneto, the source of the engine's electrical power, attached to the engine's flywheel. A Model N's magneto consisted of 16 permanent magnets. It took one worker 20 minutes to assemble all the magnets and brackets. In Sorenson's first attempt at the *lean* assembly line, he arranged 28 workers lined up along a table, each performing a single task. Ford manufacturing engineers continued to experiment with flow rates and workloads, further reducing magneto assembly time to less than 11 seconds per line, or 5 man-minutes per part—a 75% reduction in man hours. Sorensen then applied assembly line techniques to the Model T's engine, rear axle, and transmission with similar results.

Encouraged by these successes, Sorenson and other Ford manufacturing team members believed they could also apply assembly-line methods to the entire chassis. Although a tremendous undertaking, the Model T's design simplicity, plus its huge production volumes, made the large investment worthwhile.

For even more incentive, automation would enhance, not detract, from the car's quality, a principle Henry Ford refused to compromise on. Engineers had the historic motorized, moving assembly line up and running by 1914 at Ford's four-year-old Highland Park factory in Dearborn, Michigan.

After some scientific adjustments to ensure coordinated timing and efficient ergonomics, Ford engineers settled on 45 stations, each manned by 1 to 6 workers. Surviving Ford documentation describes the operation. The following excerpts are from that documentation:

- "At the first station, 1. Three men (one press man and two chassis frame handlers)"
- "…Fix four mud guard brackets, two on each side, to the chassis frame; 600 in 8 hours…"
- "At the final station, 45. One man. Drives car onto John R St. body line."

With spectacular results, the factory dropped chassis assembly time from 12 ½ hours to just over 1 ½ hours. At its peak a Model T rolled out of the factory every ten seconds.

Higher sales volumes and lower prices reflected the success of Ford's liberal use of mass-manufacturing techniques. A 1912 Model T, configured with the popular touring body type, sold for $690. Five years later, Ford produced 735,000 1917 Model Ts, over ten times the 1912 figure. Price drops reflected the advancements in automation, with the 1917 touring model selling for $360. Even after 10 years of inflation plus many upgrades, a 1927 Model T touring sedan sold for $380, while the roadster cost $360.

People appreciate Ford's Model T for its legendary utilitarian value and not so much as a stunning work of art, an honor associated with higher-priced competitors such as Duesenberg and Packard. Yet, the Tin Lizzy still radiates a subtle, elegant beauty. From an artist's perspective, the Model T exemplifies a practical form-follows-function appearance. It has nice proportions, and the body lines blend every surface and contour into a true work of art.

Henry Ford and his Model T became legendary for good reasons. Not only did the Tin Lizzy affect the rate in which America traveled in motorized vehicles, the Ford Motor Company made a major impact on American manufacturing methods. Yet, for 20 years, Ford's only offering was the Model T, and he owned the automobile market for much of its tenure. Ford's domination streak ended when General Motors came out with a more diversified product line, attracting the majority of car-buying consumers.

By the end of its production run, the Model T had become outdated. Nevertheless, Henry had a Model T sales goal in mind of 15 million cars. Production ended on May 27, 1927, a day after the 15-millionth Tin Lizzy rolled off the assembly line, a staggering number for that generation. At the time, the Model T accounted for approximately half of the world's automobiles.

Ford put America on wheels, but General Motors (GM) took the automotive industry to the forefront of the American economy. The massive corporation grew so fast during the 1920s, it became the largest producer of automobiles, as well as joining Ford as the pacesetter for advanced American manufacturing technology.

The genius behind General Motors was a successful carriage builder out of Flint, Michigan named William (Billy) Durant. Billy Durant got into the automotive industry as a parts supplier around the turn of the century. His approach to building an automobile empire was through a holding company, which he and his partners formed in 1908. They named it the General Motors Corporation. This was the same year that Henry Ford launched the Model T. Durant leveraged money from his own company and that of his partners, which included Canadian automobile manufacturer R. S. "Sam" McLaughlin (1871–1972) and a bicycle wheel manufacturer Charles Mott (1875–1973). Mott had already bought the struggling Buick automobile brand.

The new firm anchored their product line around Mott's Buick. To attract a wider range of buyers, Durant purchased several more automobile manufacturers including Oldsmobile, Cadillac, Oakland (later renamed Pontiac), and

Elmore. Other purchases for the GM family included Ewing, Carter Car, the Rapid Motor Vehicle Company (truck builder), and A.C. Spark Plug.

The company seemed to prosper, but the rapid succession of buyouts overextended Durant's credit. In 1910, the bankers terminated Durant and McLaughlin and took over General Motors. Down, but not out, Billy Durant by November 1911 had gotten back into the car building business with a new group of partners, including an American Swiss-born race car driver named Louis Chevrolet (1878–1941). Durant liked the name Chevrolet and used it for his new company's name. Along with the brand name, Durant introduced Chevrolet's famed emblem—the golden bow.

For Chevrolet, Durant followed Henry Ford's lead by targeting a mass-volume market. Louis Chevrolet did not care for the *cheap* car strategy and wanted out. Durant obliged and in 1915 he purchased Chevrolet's share of the business. By 1918, Durant owned the majority of Chevrolet as he set his sights on another, loftier goal—regain control of General Motors.

Durant wasted little time in retaking his former empire. In late 1918, the auto tycoon orchestrated a buyout with the smaller Chevrolet taking over the larger General Motors. Billy Durant kept the GM name, assigned himself president of General Motors, and then made his entry-market Chevrolet the company's lead brand.

Durant's second tenure lasted about as long as his first. Investors forced the famed founder of General Motors out for his second and final time in 1920. Pierre S. DuPont (1870–1954), of the legendary DuPont family, took over as president, but the real changes came in 1923 when Alfred B. Sloan (1875–1966) became the company's president.

Sloan introduced a few of the industry's timeless institutions, such as model year changes and customer credit, which allowed buyers to purchase a car with a modest down payment and pay off the balance on an installment plan. Henry Ford, to the contrary, did not believe in selling cars on credit. Sloan also led General Motors to pursue a more sophisticated automotive design system than his famous competitor.

Unlike Henry Ford's tight-knit team and their sketch-ideas-on-a-blackboard design strategy, General Motors established a large research center headed by engineering wizard, Charles F. Kettering (1876–1958). Kettering earned an electrical engineering degree and owned 158 patents. His most famous was for the practical electric starter motor, which replaced the notorious hand crank. The forward-thinking innovator also pursued the development of ethanol as an additive to gasoline and solar power as an energy source.

After the last Model T rolled off the assembly line, the Ford Motor

Company shut down its factories for nearly a year to prepare for the automaker's next-generation model. When the factory returned from its hiatus in June 1928, Ford began production on their new creation, the Model A. Later that year, when the Model A hit the market, Henry Ford once again had America's top-selling model. But in spite of having a best seller, the corporation lagged GM in total car sales and up-and-coming Chrysler pushed Ford into the number three spot.

# 12
# Man's Dream of Powered Flight

With the arrival of a new century, man's industrial-age innovations made great strides toward advancing transportation by land and sea, but perhaps the biggest prize of all, conquering flight, loomed just beyond the horizon. On December 17, 1903, on the sand dunes of Kill Devil Hills, Kitty Hawk, North Carolina, Wilbur Wright not only flew a heavier-than-air machine, but the Wright brothers achieved maneuverable, controlled flight. The famous voyage covered 120 feet in 12 seconds. Despite patent wars and other claims challenging their bragging rights, they were the first. Captain Thomas Baldwin (1854-1923) in the New York Times, February 28, 1914, admitted the indebtedness. "It is high time for all of us to step up and admit not a one of us ever would have got off the ground in flight if the Wrights had not unlocked the secret for us." More to the point, Wilbur noted that, "It is rather amusing, after been called fools and fakers for six or eight years, to find now that people knew exactly how to fly all the time."

The two former cyclists and bicycle mechanics, Wilbur Wright (1867–1912) and his younger brother Orville Wright (1871–1948), pulled off an incredible feat. To do so, the self-taught brothers invested in an immense amount of research, experimentation, mechanical knowhow, and creative engineering.

The brothers began their quest by studying birds and the physics of lift. Even though physicists generally understood the science of lift, the Wright brothers had to refine those traditional scientific equations. From birds, they developed a flying technique known as wing warping. The Wrights' airplanes were elastic enough that through a system of cables, levers, and straps the pilot could twist (warp) the wings to maneuver and then stabilize the aircraft along a flight path—a cyclist's version of a bird in flight. A homemade wind tunnel, built in 1901, aided in those revised calculations.

To test their designs, the Wright Brothers built both unmanned and manned gliders. Those tethered earlier models resembled kites. Manned versions required hundreds of test flights to perfect their designs. From their glider tests, the Wrights adopted their signature twin-wing design, a longer wingspan, and a hinged, front-mounted rudder. The rudder worked with the warped-wing design to make turning and other maneuvers more precise and stable.

Once the Wright Brothers felt comfortable with their manned glider, they addressed powered flight. The wind tunnel served well for optimizing the propeller design. An efficient prop was critical for flight and relied on lift principles similar to the wings.

The 1903 plane ended up weighing 750 pounds including the engine and pilot. The brothers could not find a suitable engine with the required horsepower-to-weight ratio, so they built their own four-cylinder, 12-horsepower internal combustion engine.

The historic flight at Kitty Hawk, while monumental, did not produce a practical aircraft; it did however prove that man could fly. Orville and Wilbur continued to refine their invention and finally sold their first airplanes in 1909, only 6 years after Kitty Hawk. By that time, Wright airplanes could fly distances exceeding 100 miles.

With Orville Wright at the controls and Wilbur Wright mid-stride, right, the 1903 Wright Flyer makes its first flight at Kitty Hawk, N.C., Dec. 17, 1903. *Courtesy, National Air and Space Museum, Smithsonian Institution.*

In a no-risk-brings-no-rewards business, flying the first airplanes was a dangerous occupation even for the experts. In 1910, seeking additional orders, the brothers hired a group of promotional pilots to fly Wright planes at exhibitions around the country. During the group's short one-year history, six of the nine pilots died in plane crashes. Wilbur, due to the pilots' high attrition rate from crashes, took the responsibility of instructing pilots for the firm's largest customer—the US military. Yet despite the dangers, aspiring aviation pioneers would not let a few crashes thwart man's ambition to fly. Casualties mounted, yet sales orders kept the fledgling industry afloat.

Many contemporary inventions developed at a rapid pace, but aircraft technology moved even faster. That overwhelming enthusiasm reflects humanity's eons-old deep desire to soar like the birds. From mythology to science fiction, man's imagination had already conquered flight. To adventurous aviators it must have seemed like all of civilization's scientific knowledge and clever inventions were just prerequisites to fly.

When war broke out in Europe in 1914, the aircraft industry, still in its infancy, now included American and European military forces, all rushing their aeronautical engineers to transform the airplane from a novelty into a *flying acrobatic killing machine*. Besides the United States, Germany, Great Britain, France, Italy, and Russia built military airplanes. America's most popular World

War I fighter plane was the Curtiss JN-4. Known as the Jenny, Curtiss built 6,800 of the aircraft for the war effort. Retired Jennies became famous for barnstorming acts.

After the war, civilians once again took control of the airplane industry. The bulk of World War I aircraft materials consisted of wood, varnished fabric, steel guide wires, and steel tubing to reinforce the undercarriage. Those biplanes were lightweight and compact, an ideal design for lower-horsepower engines, but their wood frames were susceptible to undetectable internal flaws and fatigue.

As engines became more powerful, manufacturers shifted design platforms from biplanes to monoplanes. Besides greater horsepower, those more modern-looking airplanes relied on the cutting-edge technology of lightweight aluminum alloys. A rigid fuselage and wing package, along with articulating rudders and wing elevators, replaced the Wright's flexible design as the preferred method to maneuver and stabilize modern aircraft.

In 1926, Henry Ford introduced the most famous of the next generation of aircraft, the Ford Trimotor. Nicknamed the *Tin Goose*, the Trimotor featured an all-aluminum alloy fuselage and wings. Per its name, three air-cooled radial-piston engines, with cylinders arranged similar to the spokes on a wheel, powered the aircraft. Its prominent corrugated, flat fuselage makes the plane easily recognizable today. From 1926 to 1933, Ford sold 199 passenger and cargo versions worldwide.

By the mid-1920s, aircraft were no longer used for just sport and dog fights. Commercial carriers, buoyed by mail delivery, began passenger service. The Ford Trimotor, with its impressive safety record, a modest five-hundred-mile cruising range, and the availability of a growing number of commercial airports, was in the first wave of airplanes that launched the big airlines such as American Airlines (1926), Pan American Airways (1927), and Trans World Airlines (1930). Airlines served many domestic routes along with a few

U. S. Army Air Mail Curtiss JN-4HM Hisso Jenny (s/n 38262, communications conversion of JN-4H) in low-level flight. *Courtesy, National Air and Space Museum, Smithsonian Institution.*

White I-30 tractor pulling Ford Trimotor airplane. *Courtesy, Wisconsin Historical Society.*

shorter international flights. US transcontinental connections still required train rides in areas with widely dispersed airports.

# 13
# Mechanized Farming

For most of history, farming chores have been slow-moving, back-breaking work. Europe's post-medieval Agricultural Revolution offered minimal relief, and the Industrial Revolution a bit more. By the mid-nineteenth century, the pace to modernize farming sped up. Two early American contributions included the steel plow (1837) by John Deere (1804–1886), and the mechanical reaper (1840s) by Cyrus McCormick. Later in the century, the combine and mechanical tillers made their impact on agricultural production. Yet, even toward the end of the 1800s, small farms still depended on beasts of burden and simple tools to assist with their chores. On the other hand, larger farms relied more on mechanization, particularly in the western plains. Ambitious farmers employed teams of up to 25 horses to tow complex machinery such as reapers. Due to farm's mechanical marvels the average farm acreage increased while the number of farm laborers decreased.

Compared to other industries, applying the latest mechanical horsepower to agriculture proved a challenge. Factories abandoned animal power for water wheels and windmills before the 1700s. Steam-engine power plants made their appearance toward the end of the eighteenth century and then became commonplace during the middle of the nineteenth century. Steam-powered locomotives replaced horse-drawn rail cars in the 1830s, then prominently went on to define the American Industrial Revolution. The introduction of the automobile quickly made the horse-drawn carriage obsolete. The machines that would revolutionize farming, tractors, known in the early days as pulling, or tractive machines, would take more time.

Similar to railroads, the first cutting-edge farmers used steam engines to pull machinery using stationary tractive motors. They positioned the engine on either side of a field and used cables to drag a plow back and forth. A wheeled tractor made a lot more sense, but what would it look like? When manufacturers finally began building tractors, the massive machines looked much like steam locomotives. Those huge iron horses, weighing in at about 20,000 pounds, are impressive today for antique tractor collectors, but the farm behemoths failed to revolutionize farming.

The wheeled farming-technology lag was not due to a lack of effort. Plenty of inventors took a shot at an engine-driven traction device. In 1906, over 180 manufacturers made tractors, both steam and gasoline powered. Agriculture though was not a capital-intensive trade. By farming's nature, "low entry barrier costs" encouraged poverty and discouraged investment. To be successful, farm tractors needed to be reliable, practical, and most of all, affordable—all traits attributed to mass production.

By World War I, beasts of burden still supplied the bulk of the tractive force to pull heavy farm machines, but the internal combustion engine was about to revolutionize farming. Gasoline power allowed tractor builders to make their machines smaller and more affordable. Some early gas-engine tractors had three steel wheels—two for traction and one for steering. Another popular tractor configuration used a crawler tread, similar to military tanks. The crawler treads had great traction, and their large footprint spread out the machine's heavy load, but they were expensive and high maintenance. The future look of tractors finally emerged during World War I courtesy of a well-recognized name.

In 1917, Henry Ford, with his son Edsel (1893–1943), launched a separate venture—manufacturing farm tractors. He named the company Fordson. Ford formed the company as a ruse to gain control over the Ford Motor Company. Nevertheless, Henry Ford always the entrepreneur took the business seriously as his team set out to design and build the Model T of farm tractors. Fordson used the now-familiar tractor-styled frame in which the engine and drivetrain form the structure of the tractor's backbone. A four-cylinder gasoline engine powered it and two massive iron wheels provided traction and supported the often-heavy towing load. A couple of wheels up front added stability and steering.

drawn grain binder converted to be pulled by a Fordson tractor. *Courtesy Fillmore County Historical Society*

Social critics, already concerned about the outlook of farming, feared that tractors would make the small farmer obsolete. However, handy tractors, the name for early small farm tractors, slashed the time to perform farming chores regardless of the size of the farm. Thus, the small farm did not disappear, but on average, the size of farms continued to increase.

At its peak, Ford supplied 75% of US tractors and 50% of tractors worldwide. The one-ton, $400 machines dramatically reduced manual labor and increased productivity. For example, perched up high and bouncing along on the steel seat, wheat farmers could pull machinery capable of plowing, sowing, harrowing, hauling, thrashing, and winnowing. For wheat, they could accomplish those tasks in half the time compared to pulling equipment with horses. Farm tractors produced similar results for many crops.

Tractors served as much more than tractive machines. For example, farmers used tractors as stationary power plants. To convert a tractor into a power generator, the farmer jacked up the rear wheels and attached a flat belt between the drive wheel on one end and a pulley on the other end which attached to whatever contraption the creative farmer desired to rotate. During its tenure, from 1917 through 1928, Fordson sold more than 500,000 tractors.

To increase a tractor's versatility, International Harvester offered the first PTOs (power takeoff) as a tractor option in 1922. A PTO drive provided auxiliary power to towed machinery through a low-speed output shaft.

Other tractor manufacturers followed the familiar, classic design of the Fordson, and the farm tractor industry began to grow. Between 1925 and 1930, sales of gas-powered tractors averaged about 175,000 per year. By the end of the 1920s, over one-million tractors worked American's farmland. Despite the great leap forward in farming technology, the purchase cost combined with the operating expenses of tractor ownership still prohibited many small farmers from using the new technology. For the time being, work horses continued to outnumber tractors, but the 1920s farm tractor made its positive impact on American agricultural production.

# 14
# Freedom through Communications

Advancements in communications rank as some of mankind's greatest inventions; a number of them revolutionized society. The inventions of some notable inventors include:

- Johann Gutenberg (1398–1468)—the movable-type printing press (1455). The invention that helped build modern Europe.
- Louis Daguerre (1787–1851)—camera (1837). "A picture is worth a thousand words."
- Samuel Morse—the telegraph (1862), with its language of dashes and dots.
- Alexander Graham Bell (1847–1922)—the telephone (1879), and his famous inaugural phone call, "Mr. Watson. Come here. I want to see you."

Communication industries advanced those technologies throughout the new century; however, the early twentieth century produced its own landmark technology—electronic wireless telecommunications.

An Italian born into an affluent family, Guglielmo Marconi (1874–1937) took an early interest in the science of wireless telegraphy. In the early 1890s, the discipline of wireless technology was only in the theoretical stage. Marconi's goal was to pursue radio broadcast, or shipping's answer to the telegraph. After all, ship-to-ship and ship-to-shore communication would be of enormous value to the maritime industry. To polish his craft, a determined young Marconi gained expertise in wireless communication through a combination of education and experimentation with small-scale models. Initially, his father funded Marconi's projects, but as the size and scope of his models increased, he needed to attract the interest of large-scale investors. In 1896, Marconi moved to England to gain better access to potential investors (Marconi's mother was British, and he spoke fluent English). The plan worked and a string of successes kept his wireless-technology projects well-funded.

Then, in 1902, Marconi made his historic, defining breakthrough. The 28-year-old inventor and entrepreneur broadcast a faint Morse code message from Nova Scotia to England. Despite his achievements, Marconi's detractors suspected the inventor rigged his tests, but he eventually proved his wireless technology was genuine.

On April 15, 1912, the *RMS Titanic* struck an iceberg. A Marconi-employed radioman stationed on the *Titanic* radioed SOS distress messages from the sinking ship. Thanks to the new technology, critical help arrived in time to save hundreds of lives. It was a great start for wireless radio, but it was only the beginning.

Marconi may have been on a quest for wireless telegraphy with an interest in shipping, but radio communication proved just as valuable for aircraft. Using short-wave radio, engineers developed both ground-to-plane and plane-to-plane communication systems. Across the country, the number of airports increased the density of radio transmission towers to the point where pilots could fly across many areas of the country while maintaining continuous radio contact.

ⵛⵛⵛ

Cutting-edge electronics were critical to the advancement of wireless and wired communications. For the telephone industry, electronics addressed the problem of too many phone calls clogging switchboards. Customer demands tasked Ma Bell with an ever-increasing number of domestic calls, plus undersea transcontinental phone cables now connected North America to both Europe and Asia. By the 1920s, the forty-something-year-old labor-intensive telephone command centers needed serious upgrades to prevent the monstrosities from being overwhelmed with heavy traffic. The solution was electronic-based automatic switch boards, which started to replace manual systems after World War I. The complete conversion to full automation proved a long, arduous process spanning decades.

The single most important device responsible for the early electronics was the thermionic valve, better known as a vacuum tube. Vacuum tubes are a class of electronic devices similar in construction to an incandescent light bulb. They control, condition, or amplify an electrical current. Scientists and engineers spent many years researching the vacuum tube but there were few applications for the technology. Then came the emergence of wired and wireless communications.

Marconi and Alexander Bell only focused on point-to-point communication, but as scientists and engineers unraveled the secrets of long-range wireless transmission, the broadcasting of radio waves through omnidirectional radiation emerged. On both ends, receiving and transmitting, the thermionic valve would be critical to the next big development in communications.

Inventors combined advanced electronics with existing wireless technology to produce the ultimate mass-media format—AM radio. America's first AM broadcast occurred in 1906. After some crucial developmental work, the radio industry took off in 1921. The end result was no small achievement. Consider that by the 1920s, engineers learned to take a faint signal from a broadcast tower, modulate it, magnify it with minimal distortion, and then blast near-live sound quality through a speaker. A timeline of music played over the airwaves suggests the escalating sophistication of the electronics. Early radio music relied on the booming sound of orchestras. Eventually, with enough advanced technology, a listener could enjoy the delicate melodies of a string quartet.

As a testament to its tremendous value, the number of radio stations and listeners quickly escalated. In its inaugural year, 508 stations began operations, causing politicians to answer the increasing need for bandwidth with the Radio Act of 1927. By the end of the decade, radios became household necessities as nearly 12 million American families owned at least one radio.

Television would not be a significant factor until the 1950s, but in 1929, the engineers at General Electric could broadcast moving pictures onto TV screens between Schenectady, New York and New York City.

The field of communication provided a tremendous boost to America's wealth providing countless hours of entertainment, delivering valuable services, and accelerating the spreading of knowledge. In addition, in an era of burgeoning democracies, it is worth noting that more than any other industry, communications hold a special bond to liberty. Gutenberg's printing operation in the 1460s became Europe's primary catalyst leading to the formation of the modern, democratic state. Printing brought a wide range of political opinions and knowledge to ordinary citizens. Radio broadcasts did the same, often breaking the propaganda monopoly spread by despots.

## 15

# The Electrical Power Grid

Early twentieth century technologies such as automobiles, airplanes, and radios transformed the American economy and its culture. Nonetheless, as influential as those engineering marvels may have been, they were not the predominant game-changers of the period. That honor went to the nation's emerging electrical grid, a vast complex of massive power stations constructed to deliver unprecedented amounts of power through a network of cables and transformers, conveniently delivered to municipalities, homes, and businesses.

The study of electricity had some of its earliest roots in the United States when Benjamin Franklin (1706-1790) made his risky experiment during a violent rainstorm. Across the Atlantic, scientists explored more practical uses for electricity, first taking aim at the development of a battery, a term created by Franklin. Progress though was slow. It took much of the nineteenth century to move from an early battery, the voltaic pile (circa 1800), to the modern dry cell battery. By that time, scientists understood electricity and engineers had developed practical uses for the novel power source. Despite electricity's potential, the outlook for an electric future appeared bleak due to the pessimistic conventional wisdom of the day. In the meantime, an ambitious inventor working out of his New Jersey laboratory was about to prove the skeptics wrong.

In 1879, Thomas Edison patented his improved incandescent light bulb. But for his invention to prosper, potential customers first needed a source of cheap, reliable electricity. Edison took on the task himself, and in 1882 he pioneered America's earliest large-scale electrical power network. At that point, large scale meant that stations, such as the one on Pearl Street in New York City, served a few blocks. Similar to a battery, Edison's system relied on direct current (DC). Over the next five years, Edison's company added 120 more DC systems. Besides powering an illumination alternative to kerosene lamps, Edison's DC networks offered the first large-scale use for electric motors.

Edison's 110-volt DC electrical grid achieved mixed success. One huge positive was that he planted the concept of a national power grid. To this day, America depends on the great inventor's centralized, distributed power concept (at approximately 110 volts) for low-power applications, such as lights and appliances. However, Edison chose the wrong electrical system.

As Edison established a foothold in the industry, a new competitive technology arose, in part from Edison's own lab. The new system used alternating current (AC). AC power grids offered greater efficiency and flexibility compared to DC grids. George Westinghouse, an inventor already wealthy from his successful pneumatic train brake, took on the challenge to dethrone the DC grid and replace it with AC. Edison once said, "I shall make electricity so cheap that only the rich can afford to burn candles," yet Westinghouse offered the better option for commercial electricity.

Practical AC required more sophisticated equipment than DC, such as alternators, transformers, and polyphase motors. At the time, that equipment did not exist. With help from talented engineers, including Edison's former employee a Serbian-born American immigrant and famous tech-genius Nicola Tesla, the Westinghouse group developed the core machinery needed for an operational AC grid.

Alternating current called for higher transmission voltages and high voltages frightened Thomas Edison. Edison believed AC to be far more dangerous than DC, plus he needed to save his own venture. In response, the Wizard of Menlo Park initiated a campaign to discredit AC power. The battle by Edison against Westinghouse to make DC the national standard of power grids became known as the war of the currents.

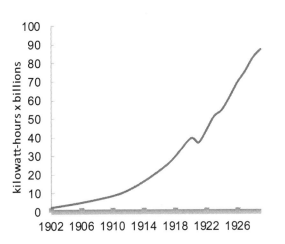

**Figure 15.1** The rapid growth of the American power grid. Electrical utilities output, 1902-1929

The conflict escalated when Edison mounted a vicious propaganda campaign against Westinghouse. For example, Edison sought to have the term electrocution replaced with the term *Westinghoused*. His technicians publicly electrocuted animals with AC power. He lobbied state legislatures to ban his competitor's product. Edison's most grievous stunt was when he convinced the state of New York to execute death row prisoners by the electric chair using AC power. The state agreed and on August 6, 1890, executed a man in an AC-powered electric chair. The execution did not go well. The large dose failed to kill the man on the first try. A reporter who witnessed the event described it as "an awful spectacle, far worse than hanging." The great inventor lost the war of the currents.

Initially, Westinghouse only used a simpler single-phase current operating at 110 volts, Edison's standard voltage for light bulbs (incandescent bulbs worked on AC or DC). However, he soon evolved his electrical network to the more resourceful polyphase voltages, a system developed by Nicola Tesla. Interestingly, Tesla gained most of his knowledge of polyphase motors and generators, the AC version of dynamos, as an employee of Edison's lab. With the addition of polyphase power, Westinghouse's AC electrical grids gained enormous versatility, capable of powering everything from a household vacuum cleaner to a 500-horsepower industrial machine.

In 1886, Westinghouse built his first hydroelectric system to power streetlamps in Great Barrington, Massachusetts. Five years later, Edison's nemesis demonstrated his polyphase system at the International Electro-Technical Exhibition of 1891 in Germany. From then on, the big contracts kept on coming, firmly placing AC as the power format of the future.

In 1892, a group of investors led by J.P. Morgan merged the Edison General Electric Company and the Thomson-Houston Electric Company to form General Electric (GE). The new leadership relieved Thomas Edison of any corporate authority. The firm also hired their own formidable AC electrical engineer genius, another immigrant named Charles Proteus Steinmetz (1865–1923).

By the 1890s, Westinghouse, soon joined by General Electric, began the immense task of electrifying the entire nation. The rapid spread of electricity reflected its tremendous utility. The invisible power source reached 90% of American homes by 1930. The electrification of rural America had to wait a few more years to catch up, but by the end of the 1920s, thanks to Westinghouse and General Electric, the majority of the nation's urban population had access to plentiful, inexpensive, and convenient electrical power—see Figure 15.1.

The AC electrical power grid featured natural, large economies-of-scale characteristics, a huge benefit. For example, a generator's design permitted larger machines (upward scalability), which increased efficiencies and lowered dollars-per-kilowatt-hour costs. Through electrical transformers, alternating current produced the high voltages (over 100,000 volts) critical for long-distance transmission covering hundreds of miles. Transformers also lowered the voltage at the point of consumption for greater safety. This dramatic increase or decrease of voltages occurred with minimal energy losses. Pure DC power does not have those traits, a limitation that proved a technical showstopper for the ambitious Edison.

To be successful, the electrical grid needed a large power source for its generators. Fortuitously, soon after Westinghouse launched his utility company, English inventor, Sir Charles Parsons (1854-1931) began manufacturing steam turbines in Europe for high-horsepower shipping and electrical-power-generation applications. Westinghouse purchased the rights to Parsons' design. Then, in typical American fashion, the industrialist enlarged and beefed up Parsons' design to accommodate the rapidly expanding American market. Westinghouse installed their first steam turbine in 1903 in Newport, RI.

In large-duty applications, the steam turbine gained a significant advantage over the piston steam engine due to its higher horsepower-to-weight ratio. As with the piston version, the steam turbine maintained its fuel versatility via boilers. Thus, the turbine's wide range of practical energy sources included coal and oil as well as the up-and-coming natural gas.

The more electrical power technology advanced into the twentieth century, the more consumers demanded the product. The steam turbine's inherent bigger-is-better design qualities allowed it to meet those exponential economic challenges. In 1900, larger units produced an impressive 1,600 horsepower (HP). That figure reached 6,500 HP by 1907, and an imposing 70,000 HP (59,000 kW) by 1930.

A technological cousin of the steam turbine, the water turbine, which shared similar technologies, also contributed to the development of the nation's electrical grid. Though during the early 1900s, fossil fuels (mainly coal) provided the bulk of American electricity, dammed rivers also produced notable amounts of electrical power. By the 1880s, a similar timeline to the steam turbine, European-designed water turbines had reached efficiencies of 90%.

Electric utilities redefined the ultimate American home. In 1920, a utility company placed an advertisement in a local paper titled *Electrify Your Home*. The ad suggested that for a 1920s residence to be modern it should have most of the following items: a sewing machine, a burglar alarm, an electric waffle maker, a vacuum cleaner, an iron, an electric water heater, a doorbell, an electric oven, an electric shaving mug, porch lights, an electric toaster, an electric percolator, and of course, a washing machine. Electricity powered an even greater variety of devices for stores and for all types of factories.

The abundance of convenient and versatile electric power set into motion a flurry of economic activity proliferating the sophisticated mass-production methods encompassing thousands of products. How the Great Depression followed this technological revolution demands a better explanation.

# 16
# Electric Motors & Ball Bearings

One of the great early electrical pioneers, Michael Faraday (1791–1867), first experimented with "electromagnetic rotation" in 1821. Faraday's research led to two similarly constructed devices, the dynamo and its inverse, the electric motor. The dynamo generates electricity from a rotating power source. The electric motor generates mechanical rotation from electricity. German and English engineers worked independently to advance the utility of both dynamos and electric motors. Larger, practical versions of the two machines became viable for commercial use by the 1870s.

Dynamos and motors suited Edison's plan to build a series of electrical grids. Steam powered dynamos supplied the electricity for his electric light bulb, while motors augmented the demand for electricity created by the light bulb.

With the introduction of power grids, and no longer dependent on expensive batteries, larger-scale electric motors made the transition from the lab to practical applications. Trolley cars proved one early valuable use for DC electric motors. Steam engines created enormous amounts of air and noise pollution, while the quiet, electric alternative ran practically pollution free. Electric motors for elevators, an essential addition to steel-structured skyscrapers, proved to be another electric power opportunity.

The emergence of electric motors gave a huge boost to society's ability to harness electrical energy efficiently. They filled a critical power-transmission void created by the limitations inherent to steam power and internal combustion engines. This new technology offered ground-breaking versatility, convenience, simplicity, scalability, plus quiet operation, and once designers overcame efficiency problems, cost effectiveness. From 1900 into the 1920s, further advancements in the electric motor made it one of the unsung technological heroes of the early twentieth century. One of the technologies that elevated electric motors to a higher technical plane was the ball bearing.

The ball bearing solved many of those inefficiency problems plaguing rotary machinery through their greater precision and lower friction. Leonardo Di Vinci (1452–1519) sketched a concept design for a ball bearing 500 years ago. Di Vinci's configuration used a series of balls constrained in a curved track that produced a smooth rolling motion similar to a wheel. The reason that technologists took five centuries to build on Di Vinci's concept is that the practical ball bearing presents numerous challenges. It requires nearly perfect spheres of equal size that must be made from a hardened material. The tracks in which the balls spin must have excellent curvature and a similar hardened material.

Not until the late nineteenth century did industries offer the appropriate steels and fabricating technology required to manufacture a reliable ball bearing. Friedrich Fischer (1849-1899) of Germany, produced the first ball bearings in the 1880s. In 1905, his company registered as Fischers Aktein-Gesellschaft (FAG). By the onset of WW I, a few American companies manufactured high-quality ball bearings. To further improve a ball bearing's performance, industries developed a companion technology—friction-reducing, petroleum-based lubricants.

The ball bearing enhanced the performance of electric motors by reducing friction while allowing for tighter tolerances, which increased a motor's electro-magnetic efficiencies. The largest gains came with smaller motors. Manufacturers of factory machines, appliances, and many other devices packaged those powerful, reliable motors into their designs. Those substantial gains came after the turn of the twentieth century. For example, the same-sized motor that produced 7-1/2 horsepower in 1897 yielded 100 horsepower by the 1920s.

The electric motor is one example of wasted energy due to friction reduced by ball bearings. The transportation, appliance, and manufacturing industries created many more opportunities.

Cut-away view of a ball bearing. *Courtesy Grayhawk Press.*

Ball bearings were the earliest in a class of bearings called rolling element bearings. The ball bearing is the simplest type of bearing to manufacture and relatively forgiving when it comes to misalignment and contamination. However, they best serve only light-duty and medium-duty applications. Large equipment, including power-plants, railroads, construction equipment, heavy trucks, steel mills, and large motors, needed a heavy-duty bearing. The perfect rolling-element bearing for bigger jobs came from an American, Henry Timken (1831–1909). Introduced in 1901, Timken termed his invention a tapered roller bearing. Even today, a tapered roller bearing requires an extraordinary amount of technology to manufacture, highlighting the magnitude of the feat at the turn of the century. Henry Timken designed his novel bearing for wagon wheels, but they soon became standard equipment for other demanding jobs.

The differences between the old-standard, a sleeved bushing, and Timken's tapered roller bearing were significant, especially when moving a massive load such as a train from a standstill. In an old advertisement, the Timken Company claimed that a typical train with its engine and loaded cars equipped with roller bearings required only one-eighth the horsepower of traditional bushed axles. Industrialists eventually realized the tapered roller bearing's value and used it for many other large rotating spindles.

Twentieth-century American industries generated prodigious amounts of power. The widespread use of low-friction bearings installed in those mechanisms transferred power more efficiently thus moving us another step closer toward a modern economy.

Dissembled tapered roller bearing. *Courtesy Grayhawk Press.*

# 17
# Precision Grinding

The modernization of technological marvels such as automobiles, tractors, airplanes, and bearings required an array of sophisticated manufacturing tools. One of those industrial building blocks, which emerged just before the Great Depression, was precision grinding machines.

A typical grinding process uses an abrasive cutting wheel, also known as a stone, to remove small volumes of hard materials. Grinding by large diameter stones cut from sandstone dates back to antiquity. Sandstone's mineral matrix contains the natural abrasive silica. A disc-shaped sandstone mounted on a simple axle and operated by a foot treadle became a necessity for farmers, woodsmen, butchers, and others for sharpening dull tools such as axes.

As industrialization advanced, machine components used materials that were harder and thus tougher to process, so manufacturers added modern versions of old-school grind wheels to their machine tool repertoire. These early grinders, circa 1890s, resembled traditional machine tools such as lathes and milling machines. For the abrasive stones, material scientists created manmade wheels using natural emery or corundum as the abrasive and a cemented, gum-resin shellac as the binder.

Twentieth century technology brought additional challenges for conventional manufacturing. For one, designers squeezed even more horsepower into smaller, higher-speed, lighter-weight packages. Then there was the sheer volume of parts. Whereas a century earlier Whitney's military musket's lock, stock, and barrel consisted of about 50 components, Ford used 5,000 parts to build the Model T. By the 1920s, annual automobile sales had reached into the millions compared to a musket's thousands. Those high-performance automobiles had to be both durable and affordable. Precision parts required high accuracy. Fast moving parts needed to run at high efficiencies. For cost effectiveness, mating parts depended on high precision for ease of assembly. Other parts used exotic materials designed to withstand harsh environments and heavy-duty cycles. To meet these extraordinary demands, automotive manufacturing engineers needed to elevate their game.

The complexity of an automobile, its reliance on factory automation, and their huge production volumes created a large infusion of cash for the machine tool industry to help meet the industry's stringent criteria for continuous improvement. This need led to a new class of high-performance grinding machines designed for the high-volume automotive component market. Other industries such as aircraft production, power transmissions, and bearings manufacturing also benefited from precision grinding.

Toward that goal, machine tool builders during the 1920s introduced new generations of specific-purpose grinding machines—designed from the ground up for precision work. For example, those machines required a solid foundation, thus grinders came equipped with extra-heavy cast iron bases. Each of a grinding machine's sliding mechanisms were capable of small, precise moves. High-speed wheel spindles handled severe conditions with minimal vibration. Innovative tooling firmly and precisely rotated the parts.

The key to those machines' higher performance was new advanced grinding wheel technologies. A recipe of synthetic abrasives bonded into a substrate and then cooked at a high temperature in electric furnaces replaced stones using natural abrasives. Those new-age materials, which included aluminum oxide and silicon nitride, produced a harder more precise grinding wheel. The next generation of wheels wore slowly and were reasonably priced, a perfect solution for mass production.

One of the extraordinary grinding machines from the 1920s was the centerless grinder. Centerless grinders featured a wide wheel and through-feed motion, a type of mass-production where parts flowed like a river compared to the more common time-consuming method of discrete movements—the equivalent of laying bricks. Cincinnati Milling Machine Company built the most popular centerless machines beginning in 1924. The company's corporate history notes that because of its centerless grinders, engine valve tappet production rose from 90 parts per hour in 1920 to 1,350 parts per hour in 1929. In another 1920s example, their centerless grinders increased the precision of engine piston pin diameters by a factor of five, which also included significant increases in the production rate. Those momentous improvements provide valuable insights into just how automobiles of the time had increased performance coupled with reduced prices.

The latest class of precision grinding machines elevated technical performance and lowered costs. Besides sharpening operations, modern machine grinding expanded its range of applications to include reducing distortion errors and making parts smoother, flatter, or rounder. Although originating in the automotive industry, those standards for high-tolerance parts became known as aircraft precision.

# 18

# The Evolution of the Modern Factory

Shortly after the turn of the century, inside the walls of American factories, electricity launched a dramatic economic revolution. The makeover transformed the methods for all of American manufacturing, from small machine shops to large mills. The motorized electric factory evolution produced a large assortment of improved, less expensive products, made in higher volumes, while increasing industrial efficiency. Those developments translated to greater wealth and economic wellbeing for the American people. As with other technologies, twentieth-century factory designs advanced at an exponential growth rate.

From the eighteenth century into the early twentieth century, factories evolved their main power source. This progression moved from waterwheels driven by fast-flowing streams or rivers to low-pressure, coal-burning steam engines to high-pressure, coal-burning steam engines and finally, to a steam turbine/generator package fueled by coal or natural gas coupled with a large electric motor—first by DC and later by polyphase AC. Each of those platforms produced more horsepower with greater efficiency. However, during those 100-plus years of evolution, the method of transferring that power from prime mover (the power source) to production machinery gained only modest improvements. A radical change arrived during the twentieth century, with the bulk of the conversion occurring throughout the 1920s, when a dramatic nationwide transformation reinvented the methods by which every factory in the nation operated.

Traditional methods of factory power transmission used a drive system known as line shafts (see diagram in Chapter 7). Line shafts employed a complex matrix of shafts, gears, pulleys, belts, and clutches, along with an assortment of supporting hardware, designed to transmit rotary motion from the factory's waterwheel, engine, or motor to every machine throughout the factory. In a typical application, a vertical drive shaft, connected to the power source located in the building's basement, ascended to the top story of the factory. That main shaft, through an arrangement of gears turned a series of ceiling-mounted horizontal countershafts extending left and right and hanging high above the factory floor. Machine lines on each floor were aligned and oriented underneath the overhead countershafts. Long flat belts, attached between the drive pulley on the countershaft to a driven pulley located on a machine spindle powered each machine. Machines with rotating shafts were arranged in a parallel, horizontal configuration. Machines with vertical or perpendicular shafts needed a simple looking, yet finicky, quarter-turn belt drive.

Several turn-of-the-century upgrades preceded the conversion to a total-electric factory. The first was the enhancement of shaft supports from plain bushings to ball bearings, a revision which reduced friction. The second upgrade substituted the main power source with a series of large electric motors from 100 to 200 horsepower, each positioned to drive groups of machinery. Builders often suspended those motors from the ceiling to gain floor space.

By the first decade of the twentieth century, factories and mills experimented with machinery driven by direct-drive motors. The electric grid supplied power to machines with their own motors and controls, much like the electrical system in a house

The transition from line shafts to the all-electric factory would not be

Machine shop with line-shaft driven machinery circa 1910. *Courtesy Wisconsin Historical Society*

immediate. To become the industry standard, direct-drive motors required advancements in four areas. First, the electrical grid required further expansion and upgrades, in other words, factories needed plentiful, dependable, and cheap electricity. Second, low-horsepower motors needed to operate with greater efficiencies and at lower costs. In the early days, motors suffered with scalability issues. Larger motors were far more efficient than smaller motors. Third, machine builders had to redesign their machines to accept direct-drive motors. Finally, industry, from builders to end users, had to have enough faith in their business climate to make the large investment in the new technology. By the end of World War One, industry had met the first three technical requirements, although economic justification lingered.

The complete conversion from line shaft to direct electric drives turned into a work-in-progress that would not be completed for decades. Relevant to Great Depression history is that manufacturers of all sizes across the entire American industrial complex completed much of the modern-factory makeover during the 1920s and a large chunk of the holdouts by the onset of World War II. Also, pertinent to the era's history, the benefits of the electrified factory were massive. Line-shaft drives had served their purpose for an early industrialized nation, but the monstrous rat's nest became a burden for modern mass production. This quiet revolution dramatically improved factory productivity while forming a more user-friendly environment for factory workers.

The ultimate factory rationalization project, electrification freed the factory from strict constraints such as machine-layout inflexibility, low equipment density, and limited machine versatility. Industrial architects for over a century had designed mechanical factories around those shortcomings. For example, line shafts had practical limitations on length, so building designs, instead of spreading outward, went up to three or four floors or more. Building and equipment planners organized machine layouts not for ideal efficiency but to align with the various line shafts. Machines needed to be configured to each other with drive shafts parallel and spaced according to pulley drop locations. Limited by the power output of the main engine, engineers parsed out precious power based on the torque and power requirements of each line. Often the plant layout conflicted with the ideal manufacturing flow of the product. Altogether, line shaft methods presented many obstacles toward an efficient, *lean* factory layout, the type Henry Ford sought, and everyone wanted to copy.

Even the line-shaft factory building served as much more than a structure erected to protect man and machine from the elements. Architects designed the building as a machine frame. The many supports to hold the shaft hangers needed to be strong, rigid, and accurately aligned. In contrast, direct-drive factories, except for crane supports, served mainly as shelters.

Free of their line-shaft constraints, modern factories had enormous advantages. Industries with motorized machines built bigger factories with larger footprints. Manufacturing engineers then arranged machines for ideal flow, not to fit the prearranged line shafting. Free from strict alignment constraints, the number of machines per square foot of factory floor increased (machine density). Electric motors also made the factory more flexible. Adding, subtracting, or moving machines became routine. If factory expansion exceeded the building's original power requirements, the electric utility company could provide more power. Overhead cranes and material handling systems moved into spaces once occupied by the jungle of power transmission devices that had stretched from floor to ceiling. Slender columns supported a roof instead of the next floor which allowed for better natural lighting. Reinforced concrete, another new technology, permitted factory floors to handle the weight and vibrations of heavier and more powerful machinery.

Line shafting had other disadvantages compared to discrete motorized machinery. Line shafts' complexity caused significant maintenance issues. Mechanics needed to pay constant attention to the system, making sure that all components were in proper working order. Yet, despite mechanics' best efforts, shafts vibrated, bushings failed, pulleys wore, and belts broke. The replacement of any part in a section required the shutdown of multiple machines, sometimes an entire line. Excessive downtime became an expensive, unfortunate way of life.

The openness of rotating parts with line shafting created a major safety liability. Overwhelmed by the size and th number of components in a factory, mechanics found it impossible to protect people properly from the web of moving parts. Workers needed to be on their guard not to get their clothing, hair, or hands caught in rotating machinery. Without guarding, a busted belt became a lethal weapon that covered a lot of ground. Then there was the lubricating oil dripping from bushings and bearings that created slippery floors.

Direct electric motor drives also transformed equipment industries. Machine builders introduced new generations of machinery that were faster, more powerful, and had greater agility for cutting, forming, moving, assembling, and inspecting operations. Variable-speed motors and hydraulic control systems offered flexible feeds and speeds, adding more tools with which to hone factory processes. The ease of installing multiple motors for a variety of tasks on a single machine increased versatility. Some of those new machines performed the same functions previously requiring two or more machines. Finally, direct-drive motors delivered more power with greater efficiency. Henry Ford, the industrial pioneer who took the most advantage of the power grid and early electric motors, acknowledged that the revolutionary Model T would not have reached its level of production without the advancements made to the industrial electric motor.

Sweeping evolutionary changes in factory production occurred over many industries at the same time that inventors were releasing novel products, driving rapid American modernization during the early twentieth century and making mass-consumer goods a reality.

Even a rough estimate of productivity gains or cost-to-benefits due to the elimination of line shafting is impossible, but it must have had a profound and positive effect on the economy. Yet, the economic windfall contradicts conventional wisdom's view of the pre-Great Depression era and the following decade-plus long economic depression.

Industry did its job to move the nation forward, but what about its political leaders? Does the 1920s emergence of the electric factory point to a smoking gun implicating inept politicians as the cause of the Great Depression?

# Part Four
# How to Break a Robust Economy

# 19

# Socialist Ideologies Come to America

The vigorous 1920s factory culture brought America to the edge of modernization. Then, despite industry's extraordinary momentum, the economy tanked. What happened?

An aggressive version of a business-cycle recession is the primary reason for the initial plunge—the October 1929 stock market crash. To explain the extended recession, scholarship becomes more complicated; however, there are plenty of clues. For example, the onslaught of federal regulations and programs, all paid for by excessive taxes, left its trail of destruction. The total of politicians many blunders then go on to help explain the lengthy disaster that followed.

The 1920s modern industrial society encouraged the modern political society. Leading the Roaring Twenties radicals were those promoting socialist ideologies. The trend swept through the nation following World War I, shifting the center of academia's political views to the left. This new generation of *intellectuals* believed the power of the federal government could be the great equalizer against capitalism's flaws, such as social injustices, corrupt business practices, and unstable monetary systems.

Influential foreign sources of *enlightenment* came from Italy and the Soviet Union. Italy's Benito Mussolini (1883-1945) came to power in 1919. *Il Duce* gained fame for his trains running on time. The Bolsheviks founded the USSR in 1922. The Soviet's socialist agenda aimed to close the growing income gap between the rich and the poor. Back home, liberal Americans awaited the outcome of those social experiments. When the US market collapsed in 1929, progressives increased their rush to intervene, putting anti-capitalist philosophies into practice.

The labor movement gained newfound friends in Washington. Big labor knew that they needed *privileged* legal clout at the federal level to give their movement crucial parity with corporate America. The frustration of high unemployment and low wages gave a sense of urgency to the partnership.

Remnants of the latter-nineteenth century Greenback Party's beliefs morphed into Keynesian economics. Keynesians believed that increasing the number of dollars (deficit spending) produces across-the-board economic growth. Even though the popular movement had limited effect on the Great

Depression, it had tremendous influence on how today's academics analyze economies—including the Great Depression. It so happens that the left-leaning group's *creative* mathematics make private sector progress seem lackluster and bad government initiatives appear positive. Was the popularity of their theories a result of brilliance or good luck?

There was also plenty of retro politics. Much of the idealism that caused the United States economy to spiral downward had its heyday dating back to Europe's enlightened era 300 years earlier—a system known as mercantilism. For the Great Depression, post-medieval zombies arose from the dead, putting the brakes on economic growth via the archaic practice of mandating higher prices. Although low prices fueled the 1920s expansion, social leaders made convincing arguments that raising prices would stimulate the 1930s economy.

Politicians and bureaucrats packaged their high-wage, easy-money, radical-labor, and socialist doctrines into a New Deal, all designed to rejuvenate the ailing economy and revamp society. Optimistic and energetic, liberal idealists' plans overwhelmed traditional capitalism. Holding to a belief that the US Constitution should not stand in the way of a sensible recovery plan; their actions pushed the economy deeper into despair. Rather than recognize their faults, they doubled down their big-government efforts. As political activists, this motley collection of enthusiastic revolutionaries emboldened federal politicians, judges, and bureaucrats to join their cause.

Those social experiments produced a historic enigma, one far more devious than realized by today's establishment historians. Although the grand recovery effort did not reverse record-high unemployment or a lackluster business environment, incredibly large segments of American manufacturing proved more productive in the 1930s than it did during the 1920s. Understanding Hoover's and FDR's *experiments* goes a long way toward unfolding the mystery of how despite America turning modern there was the Great Depression.

By the spring of 1930, the nation became embroiled in these two powerful but conflicting movements. One, American exceptionalism, approached the 1930s with enormous momentum. The other were the New Dealers and their like-minded compatriots' efforts to revamp the nation's economic system and unintentionally place a substantial drag on industrial productivity, the topic chronicled in the following chapters.

Let the games begin.

# 20
# Déjà Vu

The Great Depression was not supposed to happen. Motivated by the Panic of 1907, Congress wanted to prevent such future catastrophes, so they established the Federal Reserve System in 1913. Seven years later, another economic disaster struck the American economy, but the Feds got a pass. Experts blamed massive European spending during World War I and not a faulty central banking system as the cause of the 1920-1921 recession. But what happened in October 1929? Was that crash beyond the control of the Fed—a result of extraordinary influences?

Ever since the decade of economic prosperity ended, pundits have expressed a wide variety of views as to the hows and whys of the economy's demise. Despite a colossal amount of academic research on the subject, eliminating many of the popular suspects only requires casual scrutiny.

At the top of the non-factor list sits the left's umbrella argument—the failure of capitalism. As noted, private enterprise flourished from 1922 through to October 1929. Not only had the 1920s US economy outperformed any previous decade in world history—it achieved the feat by a wide margin. Manufacturers of the era produced tremendous wealth, including an unprecedented share for the lower income classes. The vitality of the decade launched a cultural revolution that included a wave of consumer optimism. Most importantly, early twentieth century technology assembled a network of industrial building blocks capable of far greater future accomplishments. Today, the capitalist-failure argument continues to generate a modest amount of interest; however, academia now places the bulk of the blame for the panic's front-end downfall on the nation's financial systems. The leading suspects include private banking, the Fed, and the nation's gold reserves—arguments later refuted in Chapter 26.

Academia might have been trying too hard to find complex PhD-grade reasons for the economy's slump, when all along there has been a simple explanation for the forces behind the 1929 decline: Blame the economy's *business cycle* as the principal reason—the guilty party. Today this argument has modest bipartisan support and is backed by extensive evidence.

Business cycles originate from a combination of the fickle social side of economics combined with the truism, "what goes up must come down." Restless by nature, economies operate in a state of continuous flux with economic activity either on the increase or on the decrease. When heightened to an extreme, a purchasing frenzy creates a bubble, and bubbles burst, resulting in a recession. To describe those economic undulations, economists created the "business cycle" term during the 1920s.

Business cycles are sloppy, unlike the clean, repeatable cycles found in nature and technology. Still, they follow a general-shaped pattern as shown in Figure 20.1. Those patterns contain four distinct phases.

Beginning with a post-recession recovery, economic activity may climb for years at a modest but steady pace. Long-term optimism creates euphoria, which turns to jubilation resulting in an economic bubble. On the upside of a bubble, the unemployment rate falls as consumer purchases rise to unsustainable levels. Inevitably, overspending and excessive debt leads to a recession. That is the "what goes up must come down" element of a business cycle. Likewise, what falls will rise, which leads to a recovery. And the cycle repeats.

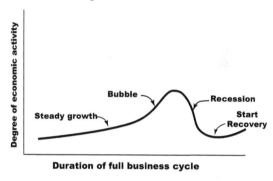

**Figure 20.1 Diagram of business cycle**

Two recessions occurred during the Great Depression—one beginning in 1929, 8 years after the 1921 recession, and the other in 1937. Those spans of 8 years are characteristic of the many business cycles which have occurred throughout American history. World War II disrupted the post-1937 cycle.

The Dow Jones Industrial Average does a magnificent job tracking a business cycle via consumer confidence. Notice how Figures 20.2 (DJIA 1924-1930) and 20.3 (DJIA 1933-1939) depict the Great Depression's two recessions. Each graph tells an important story. Figure 20.2 suggests that a classic

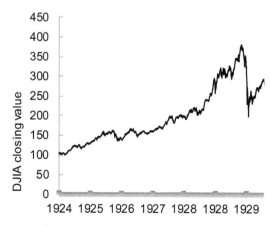

**Figure 20-2** DJIA reflects 1929 business cycle recession

**Figure 20.3** DJIA reflects 1937 business cycle recession

business-cycle episode triggered the crash which preceded the Great Depression. The other, Figure 20.3 exemplifies the power of the business cycle. Undaunted by the combined forces of an active central bank, generous federal funding, and pockets of a vigorous private sector, the economy continued its erratic ways.

The economic problem continues today, which is find a magic bullet capable of flattening a business cycle's three phases—dampen the bubble surge, lessen the collapse, and stimulate a recovery. The central bank strategy might work with Keynesian theory. In real life, conquering the business cycle is at a minimum a formidable task if not an impossible one. Human nature has its stubborn side.

# 21
# Chaos in a Robust Economy

Hard times symbolize bad economies. Businesses fail. People lose jobs. Society's most vulnerable citizens risk hopelessness. Conversely, hustle and bustle epitomize a strong economy. Factories hum. People on the move clog roadways. Masses of workers engage in exciting and lucrative careers. Shopkeepers struggle to keep pace with customers' eagerness to spend their wages on trendy gadgets. Commerce income flows while markets recycle the cash—what they do not hold for personal gain turns into investments, creating more prosperity. At least that is the enlightened view of a robust economy. Unfortunately, strong economies have their dark side.

A regrettable and yet unavoidable trait of capitalism is that during the healthiest of times economies must endure chaos—a quandary created by the same entities building economic wellbeing. So, while an economy prospers, growth industries leave a trail of destruction. This paradox applies to the strongest economic times including the 1920s, a decade that set a new high bar for economic expansion.

From certain perspectives, healthy capitalism resembles busted economies. In the wake of progress lie bankrupt businesses, unemployed workers, and devalued trade skills. The economic pie might grow larger but winners produce losers. The primary culprits are the victors, those who invest in emerging technologies and then execute an astute business plan that destroys the competition through cost cutting. Smaller businesses often find it impossible to match the bargains and will fail unless they find a niche market or negotiate a buyout.

The vibrancy of the 1920s placed the United States in that contradictory mode. Blame the popularity of the automobile, the spread of the national electrical grid, electronics, the electric motorized factory, along with the advancement of many other technologies. Through progress, the big guy seemed predestined to destroy the small entrepreneur.

The automotive industry is the classic example of "industrial survival of the fittest" and is relevant to the pre-depression era. In 1905, about 150 manufacturers produced automobiles, with new startups regularly entering the market. Within a few years, the number of automakers peaked at over 300. Then, automation took its toll on smaller producers. Henry Ford put the first major dent in the flooded industry with his popular Model T. Competition from General Motors, and to a lesser extent Chrysler, both of which offered popular models designed to appeal to a larger class of Americans, closed hundreds of automakers. Today, Ford and GM remain as the only true American car companies—three if you count foreign-owned Fiat Chrysler.

Henry Ford repeated the modernization trend with tractors. Before Ford's Fordson 180 companies-built farm tractors. Mass production reduced that number significantly. After Ford, businesses needed a factory and not a backyard shed to build a competitive tractor.

During the laundry-appliance industry's infancy following World War I, nearly 800 companies produced washing machines. As electrification drove the technology from hand cranks to electric motors, demand increased, while industry investments became capital (machinery) intensive. The cost of machinery and the need to maintain high quality with lower costs presented a relentless series of obstacles to smaller competitors. By the 1930s, only a handful of washing machine manufacturers remained.

In 1910, 500 companies in the United States produced automotive tires. That first-generation rubber was soft, lasting hundreds of miles, not tens of thousands. The combination of a growing automobile industry and quick-wearing tires assured that there was ample room for competitors within the tire industry. Then in 1912, manufacturers added a compound called carbon black. Carbon black made tires significantly tougher. It also turned the color of tires from white, the color of natural rubber, to black. The French mascot Bibendum, known in the US as the Michelin Man, shows his age by his white coloring. The additive proved great for consumers, but a disaster for the bulk of tire manufacturers.

Technology's march toward prosperity also took a tremendous toll on many of the skilled trades that built America's early industrial machine. Because of modernization, craft trades spanning from bobbin boys to wheelwrights vanished by the hundreds. As modernization spread, workers employed as general laborers and professionals often transferred into other industries. It was not always the case for those with specific skills. To survive, many proud tradesmen had to make the frustrating transition to the unskilled labor pool.

More than just statistics and anecdotes, capitalism's dark side reflects the turmoil caused by innovation. Economic volatility and bad luck crushes dreams and destroys lucrative, fulfilling careers. Safety nets might provide food and shelter, but even then, the frustration of dimmed futures might lead to mental depression and broken families. Of course, that is the case in any economy, from the productive Great Depression bookends, the 1920s and the 1950s, to the panic-stricken 1930s.

Fortunately, economic opportunities appeared to outweigh the casualties. At the top of the benefits list are the falling prices and the expanding selection of new products. For most workers, life improved. Modern economies offer steady employment along with work that is less strenuous and hazardous. The more relevant stories about commerce from the Industrial Revolution to the 1920s are not about Americans at their lowest, but the stories of their many successful adaptations. Previous chapters showcased examples of emerging companies capitalizing on new generations of technological marvels. The following cases illustrate classic corporate adaptations as they faced changing markets.

In 1855, the Torrington Company, a manufacturing firm founded in Connecticut's upper Naugatuck Valley, mass-produced industrial sewing machine needles based on a patented swaging (metal forming) process. Competition eventually cut into Torrington's market share; however, their technical expertise with sewing needles allowed them to capitalize on the 1890s bicycle craze by making wheel spokes. A few years later, the popularity of automobiles caused the collapse of the bicycle industry, so Torrington again transferred technological knowledge and moved into making heavier wheel spokes for cars. Three decades later, the introduction of formed-steel wheels slashed deep into the spoked wheels market. Yet, during the late 1920s, the 75-year-old company continued to grow as they introduced a low-friction bearing designed for automobiles and other machinery based on similar technologies used to produce sewing needles and wheel spokes. Torrington appropriately named their new product a needle bearing.

Similar opportunities repeated themselves countless times, often creating a chain reaction. The Mack brothers made the natural progression from building carriages and wagons to buses and trucks. The influx of business vehicles' created opportunities for teamsters to become truck drivers. In the construction industry, steam shovels, and later diesel excavators, replaced ditch diggers with machine operators. Not only could the shovel operator remove tons of dirt, he could work longer hours and extend the length of his career. For others out of work, the decline in manual labor provided a valuable labor source for factories. Their new careers, often assisted by labor-saving manufacturing equipment, offered a superior alternative to backbreaking manual work.

Industrialization drove a higher quality of life, but could social engineering provide an even greater lift to societal wellbeing? Progressives, the *enlightened* political faction, believed so. Formed in the late nineteenth century, their popularity climbed throughout the twentieth century. With their versatile message, progressives found acceptance across the political spectrum.

Progressives addressed social dilemmas such as poverty, racism, women's rights, alcoholism, and income disparities. Emboldened by their causes, these new-age social engineers thought they could fix some of capitalism's biggest flaws. Specifically, progressives sought to end the perpetual economic chaos, yet continue to promote industrial progress—the good without the bad.

As is often the case, do-gooders' best intentions can produce lousy, unintended consequences. For the 1930s, meddling in the massive and complex free market led to a predictable outcome. To the nation's economic detriment, commercial regulation meant applying the brakes on progress.

For the first few decades of the twentieth century, progressives' economic policies caused limited damage. But as technology moved forward and the economy sank the new establishment stepped up their efforts—collapsing an already weakened economy.

# 22

# A Return to Mercantilism

The radical mood of the times encouraged politicians to unleash a salvo of legislative bills designed to rejuvenate the sluggish Great Depression economy. The liberals (progressives) believed that unfettered capitalism was unfair to small businesses and that they could fix its shortfalls by regulating industry. The Socialist-driven labor movement argued that workers needed government-backed clout to raise wages, not just for their own benefit, but to induce more spending for the sake of general economic prosperity. There were also global influences. Italy's fascist policies impressed many American politicians. Far-left political leaders appreciated the altruistic elements of communism, particularly the communal farms of the Soviet's Five-Year plans. These various left-leaning agendas contributed to the political climate of the Great Depression. Yet the primary ideology that first sent the United States economy spiraling downwards predated modern philosophies by two centuries— mercantilism.

Mercantilism's domination of European economic-political policy spanned from the sixteenth century up to the mid-eighteenth century. Mercantilism strove to enrich the original crop of Renaissance nation states that formed throughout Europe and whose increased commerce brought an end to the Middle Ages.

Mercantile economies got a huge boost from Europe's Age of Discovery. With commerce in mind, top European mercantile nations directed state-chartered companies, such as the Virginia Company (1606), the Dutch West India Company (1621), the Massachusetts Bay Company (1629), and the French East India Company (1664) to colonize the world. Because of that global expansion, both international trade and domestic commerce flourished across Europe.

While it had great success for its time, mercantilism operated as a command economy. For example, many of the mercantile principles focused on keeping prices and wages high through regulations. To that goal, the primary economic strategies of mercantile nations included: protectionist trade policies, the hoarding of gold and silver, the establishment of government-controlled monopolies, and heavy regulation of local businesses.

In those days, a command economy came naturally. Europeans living over three centuries ago believed more in tradition and less in market competition.

Instead of survival of the fittest, they viewed commerce as a gentleman's game. Their partly written, partly unwritten business code considered a competitor as someone who needed to feed, house, and clothe their family. To them, what faithful Christian would drive their neighbor out of business to satisfy their own greed? While their aims may have appeared virtuous, their methods reduced liberty and added extra layers of counterproductive regulation.

Seventeenth-century Mercantile France offers a couple of excellent examples. Though extreme by today's standards, a lighter version of the ideology resurfaced during the Great Depression.

In mid-seventeenth century Europe, textile manufacturers developed an inexpensive weave made of unbleached, inferior-grade cotton fibers, known as calico. The low cost fabric became popular for both the poor, due to its low cost, and to the rich, as a fashion trend. But calico's success came at the expense of traditional textile mills producing higher quality cotton and other fabrics. The new competition resulted in violent skirmishes between the newcomers and established industries.

Under mercantile jurisprudence, much of western European law prohibited the manufacture and sale of products that competed against traditional technologies, especially those items in direct competition with Europe's powerful trade guilds. The regulatory problem was greatest in France. Calicos' huge popularity so outraged the traditional French textile industry that vigilante groups and law enforcement officials spent considerable time and effort confiscating the contraband—attacking or arresting offenders propagating the illegal industry. From that mass hysteria, mobs destroyed tons of fabric. Thousands of people ended up in jail. A few violators paid the ultimate penalty for their capital crimes (of trade, not violence) and were executed by the government.

The button industry faced a similar fate. During the same era, the mid-1600s, textile mills got into the business of producing buttons from wool, undercutting the price of the button guild's customary wooden buttons. For their cost-cutting offense, the government executed a few perpetrators.

Authorities considered crimes of commerce severe because they went against both the humanitarian and religious doctrines of the era. Not only did competition harm their fellow man, they robbed food from hungry mouths. Further, the novel processes disobeyed tradition, a higher power's directed methods of manufacture and sale.

By the late-eighteenth century, mercantilism faded away, replaced by various forms of emerging capitalism. The teachings of a group of enlightened philosophers took the intellectual lead of the new era. Many are still well-known today and include John Locke (1632–1704), Voltaire (1694–1778), David Hume

(1711–1776), Jean-Jacques Rousseau (1712–1778), and especially Adam Smith (1723–1790). They added to man's appreciation of economic freedom, yet their influence on a radical type of economy based on liberty (capitalism) was probably minimal. The hostile takeover of mercantilism was more due to the Industrial Revolution, which dethroned mercantilism's control. Industrialization overloaded the mercantile structure with a flood of innovation and commerce.

Under capitalism, economies grew at rates that were magnitudes faster than any mercantile economy. Whereas mercantilism demands stability and high prices, capitalism thrives on innovation, which drives down prices on established goods and provides a continuous stream of new products.

Mercantilism never vanished—though the timeline for the archaic economic system might appear decisive, but it was quite blurred. Mercantilism's meager remnants, left over from the Industrial Revolution, not only survived but flourished. In fact, those principles prospered well enough that by the 1930s they nearly took down the mightiest capitalist nation on the planet, and in a chain reaction, most of the world's other developed economies.

American politicians, from the early days of the republic on, have employed several mercantile-inspired initiatives designed to put the brakes on capitalism in the belief that if left alone, competition will destroy economic prosperity. Both Hoover and Roosevelt subscribed to many of those ideals—keep prices high and wages will follow, then the economy will flourish.

The rest of this chapter outlines an eclectic group of popular mercantile doctrines introduced into American politics and which later prospered during the Great Depression.

## Protectionism

Modern capitalism had displaced mercantilism by the time the United States of America formed as a nation in 1789. Even so, the first piece of legislation passed by Congress was a mercantile-styled tariff act spearheaded by the Secretary of the Treasury, Alexander Hamilton. The law's primary intention was to promote homegrown manufacturing through classic protectionism.

International trade, free of protectionism, is critical for any strong economy. Adam Smith made that convincing argument in his well-known book *Wealth of Nations* (1776). The bulk of economists since then have supported his treatise. Tariffs cut the volume of imports, leading to reductions in exports, thus lowering economic growth—the higher the tariff, the greater the harm. Unfortunately, the philosophers were saying one thing, political nobility was doing another.

Alexander Hamilton and his allies presented a concept they believed to be a loophole in the free-traders' arguments. They floated the presumption that in the instance of startup businesses, also known as *infant* industries, a short-term protectionist policy, in this case tariffs, should yield the long-term benefit of economic growth. Those advocates believed that infant industries needed protection from low-cost imports to survive. If it came to a price war, how could developing American industries compete against established European enterprises? While the argument may sound reasonable, the scheme has two significant flaws most notably argued by Milton Friedman.

First, with enough good PR, most companies can make a legitimate claim to infant status. Many companies have infant products struggling to gain market share. Therefore, a policy of protecting emerging products is a policy to protect all industries.

The second problem is that industries under a protective umbrella mature in a less-competitive environment. Due to an ingrained dependency on that support, there is never a good time to remove the tariffs. Taking the infant strategy to the extreme, every business can justify a protective tariff and each one of those companies, or even entire industries, will fail to reach their free-market potential.

Despite his best efforts, Hamilton made only marginal gains toward privileged tariffs. Instead, most tariffs during the nation's early decades were purpose designed, intended to raise income for the federal government. However, for populists the infant concept stuck. A few years into the nineteenth century, a long-term tariff battle erupted between the Republicans, representing northern industrialists, and the Democrats, representing southern farmers. The pattern went as follows. Republicans, when in charge, pushed for higher tariffs, and when the Democrats held the majority, they repealed them. The first shot fired initiating the tariff/free-trade cycle was the Tariff Act of 1816 during James Madison's (1751–1836) second term.

Over the next 114 years, to the detriment of the economy, American presidents signed into law more than a dozen punitive tariff acts, often leveraging high tax rates on specific foreign goods, some over 30%. The Tariff Act of 1828 during the presidency of John Quincy Adams (1767–1848), also known as the Tariff of Abominations, included some of the highest tax rates on imports in American history. Another famous protectionist bill, the Morrill Tariff of 1861 under Abraham Lincoln (1809–1865), established the new Republican Party as a supporter of high, punitive tariffs.

Thanks to free-trade Democratic president Woodrow Wilson (1856–1924), America entered the 1920s on the plus side of the protectionist trade war. But under the Republican Warren G. Harding (1865–1923), Congress crafted the Fordney-McCumber Tariff of 1922, a bill heaped with excessive punitive taxes on imports.

Regardless of the high tariffs and strong economy of the 1920s, some industries and many farmers continued to struggle. So, on the eve of the Great Depression, Congress debated on whether to legislate even higher tariffs. Those talks hinted that the new bill would not just increase tariffs but entail some of the highest and most destructive punitive taxes on imports in the nation's history.

Unfortunately, trade tariffs were not the only mercantile tool used by American politicians for attacking American industries' trend of competitive prices putting pressure on 'less-competitive' businesses.

## The ICC and the Alphabet Regulators

By the late nineteenth century, railroads had become America's largest industry, and a competitive one. Hundreds of rail carriers competed against each other for both passengers and freight. The fierce rivalries led to rate wars, chaotic pricing structures, and the frequent bankruptcies of railroad companies.

Customers were not happy either. For many of the railroads' patrons, the system appeared to be unjust and out of control. Rates constantly changed. Based on passenger volumes, railroads sometimes charged a higher price for shorter routes than for longer ones. A trip from New York to Buffalo (400 miles) might cost more than one from New York to Chicago (800 miles). Smaller businesses, struggling to survive, accused larger firms of having an unfair profit advantage because of their greater leverage which earned them volume discounts. Finally, the power of the railroad tycoons made people fear that widespread price gouging could become common practice.

A coalition of special interest groups, led by farm lobbies, pressured the United States Congress to step in and fix these *free-market problems* plaguing the railroad industry. Congress obliged, and in 1887 they established the Interstate Commerce Commission (ICC), the nation's first self-governing regulatory agency. Congress assigned an independent board to govern the ICC—free from railroad influence. In their minds, the solution supplied balance, the ideal modern political compromise—part laissez-faire and part command economy.

Backed by the power of the federal government, the ICC set out on a mission to regulate railroads for the public good. Despite the introduction of new codes, contemporary jurisprudence restrained the governing authority of the ICC. In an era in which the Supreme Court still viewed the role of the federal government as limited, the early ICC struggled to gain enough authority to enforce its initiatives. Nevertheless, as the decades passed and the progressives gained more influence in both the courts and the legislature, the ICC expanded its clout. Congress later passed several landmark laws providing more bite for the regulatory agency. They included the Elkins Act—1906, the Hepburn Act and Mann-Elkins Act—both of 1910. Finally, an empowered ICC had the means to *reform* the railroads.

The advertised primary goal of the agency was to limit excessive rates, but in classic mercantile fashion, the bureau strove to maintain big-railroad legacies by using their authority to punish those offering discount rates. For example, the ICC set prices that were more proportionate to distance, but due to the railroads' gaining influence within the regulatory agency, longer-distance rates generally increased instead of shorter-distance routes decreasing. The ICC also limited cut-rate deals, often issued through rebates, to the railroads' high-volume customers. The commission deemed them unfair to smaller businesses.

The actions of the railroad-industry bureaucracy pleased a lot of people. Congress was delighted because they intervened and succeeded at solving a problem in which, from their general viewpoint, the market system had failed. A pricing structure that made sense satisfied those citizens struggling to understand supply and demand. The railroads, although initially opposed to regulation, gained the most. The ICC reduced price competition, securing stable prices and steady work for the most influential railroads.

For the sake of economic prosperity, the ICC offered few positives, as is often the case with political agendas involving industry. Apart from Congress' best intentions, the ICC evolved into a classic fox guarding the henhouse—more of a proxy for the railroads than a public-service watchdog. In fact, under the control of the ICC, railroads prospered, but that success was not due to the ICC.

Commerce among competitors is rarely friendly. Rivalries might even have a battlefield element to them. Railroads in the decades before and after 1900 were no exception. In truth, the ICC interfered with a well-functioning competitive industry managed by a bureaucracy that succeeded in spite of itself. A couple of major factors that were outside the regulatory agency's control worked to their advantage. The first was a flood of advancing railroad technology, and the second, an all-around robust economy that depended on rail service.

From the late nineteenth century and beyond, the railroads remained a rapidly evolving industry. Durable steel rails replaced the traditional ones made from softer iron. Railroad manufacturers substituted bronze-sleeved wheel bushings with low-friction, long-life roller bearings in both locomotives and cars. Sophisticated designs made engines more efficient. Boilers ran at higher pressures and steam engines became larger, leading to significant increases in power and greater fuel efficiencies. Those changes and others meant that despite the bureaucratic interference, the railroads could reduce costs by carrying heavier loads and moving more people at higher speeds. The railroad industry innovated, and the ICC took the credit.

The surging economy kept feeding the railroads opportunities for new business. The sizzling consumer society of the Roaring Twenties meant the railroads needed to deliver more raw materials to factories and then finished goods to retailers. Even motor vehicles added to their success, at least for a while. Although during the 1920s the automobile caused a decline in the volume of rail passengers, trucks more than made up for the loss as they augmented railroad service. Local trucking, which was the extent of the industry in the 1920s, could deliver more goods to and from train terminals in much less time than teamsters and horse-drawn wagons.

Because the railroads flourished following their regulation, pundits considered the ICC a model of success. In their view, certain partnerships between the private sector and government improved the American economy. Thus, the ICC set the precedent that government should intervene in the marketplace when the widespread danger from too much or too little competition might place the public's wellbeing in jeopardy.

Toward that mindset, the responsibilities of the ICC expanded during the 1920s to include the regulation of the telegraph and telephone industries (the Federal Communications Commission (FCC) took control of their sectors in 1934). Yet once again, regulators became a tool for established big businesses to keep that status amongst reduced competition.

Once up and running, federal regulatory alphabet agencies applied their economic brand of friction throughout the Great Depression. They did more than their fair share to ensure a long and slow recovery.

## Anti-Monopoly Legislation

By the mid-nineteenth century, America had positioned itself to become the world's foremost industrial superpower—a nation captivated by man's ingenuity. Steam engines powered massive locomotives covering 30,000+ miles of track. The nation's mighty freighters steamed across the great oceans. Each day, American mills produced tons of iron. Factories hummed as labor-saving machines mass produced prodigious amounts of modern goods. Complex gadgets revolutionized farming. Yet for millions of Americans, when the sun set their active day ended. In the land of innovation, technology had failed to offer the American people practical lighting.

Candles produced only a dim light and were a fire hazard. Coal-gas lanterns emitted a pungent odor, and as the once-bountiful whale populations vanished, whale-oil lamps became cost prohibitive. The hopes for a better lamp fuel appeared in the early 1860s from a petroleum-based fuel called kerosene. Kerosene had the potential to be an ideal lighting source. Recently discovered rich oil fields centered in northwestern Pennsylvania, supplied a nearly inexhaustible source of the raw material. Plus, kerosene was inexpensive to manufacture and the lighting fuel was relatively safe and odorless.

In those early years, kerosene's high demand and the low cost of building an oil rig, transporting the fuel, and refining oil into kerosene created boom towns throughout the oil region. Competition became fierce. Price wars, sabotage, newfound fortunes, and bankruptcies were commonplace in what resembled the Wild West in the *civilized* East's own backyard.

Fortunately, some order to the chaotic oil business lurked around the corner. In 1863, a 24-year-old businessman, John D. Rockefeller bought into a startup oil refinery company. The business prospered over the next several years. Meanwhile, the young man bought out old partners and then formed new partnerships. In 1870, Rockefeller found the right combination of associates and, with his new team, established the Standard Oil Company, also known as The Standard.

John D. Rockefeller honed a business strategy like few others of his time. For one, he was not afraid to borrow money if he could make the right business case, which often included shrewd business deals. His meticulous attention to small details and his gifted business insights ensured a healthy profit. The intelligent and thrifty businessman then poured those excess revenues back into the company. Rockefeller believed in high quality. Not only was his kerosene top grade, but the same commitment to excellence applied to his buildings, production equipment, and employees. The competition did not stand a chance.

Another Rockefeller goal was to eliminate waste. As his business grew, healthy profits permitted The Standard to hire scientists and engineers who devised methods to turn waste into earnings. Whereas most refineries flushed kerosene byproducts into the river or burned them, Standard Oil reprocessed and sold them as profitable goods. The Standard's kerosene byproducts included high-quality lubricating oil, Benzene, Naphtha, and Vaseline petroleum jelly.

A fact often ignored by both its contemporary and modern detractors is that for years, to the further benefit of American prosperity, Standard Oil exported over half of its kerosene to foreign nations. Its primary global competitor, Russia, had its own substantial kerosene industry, but even in neighboring Europe, the Russians could not match Standard Oil's high quality or low prices.

At its peak in the 1890s, Standard Oil produced between 85 to 90% of all the petroleum goods sold in the United States. Since they dominated a huge market, the company, along with its chairman and major stockholder John D. Rockefeller, earned substantial fortunes. This was the age of the robber barons and Rockefeller was the most infamous.

Editorial attacks on those successful industrialists became a sport for a group of journalists known as muckrakers. The well-published assaults led by (1857–1944) and leading newspapers convinced the public that Standard Oil ran not just a monopoly but a ruthless monopoly. In 1906, the Department of Justice stepped in and charged that Standard Oil was in violation of the Sherman Act, specifically "for sustaining a monopoly by restraining interstate commerce." The DOJ hired a modest-sized army of prosecutors to make its case.

Congress had passed the Sherman Antitrust Act in 1890. The Republicans authored the Sherman Act as a political maneuver to give the impression that the pro-business party would not tolerate abusive trusts, the predecessors of corporate giants. Intended as more of a PR stunt than a true reform, the federal government kept the big-business law shelved for nearly two decades—that is until the Standard Oil indictment. From that time on, the antitrust floodgates were open.

From the start, the charges against Standard Oil and Rockefeller were a farce. For example, although he kept the title of President, Rockefeller's involvement in the business had ended 10 years earlier. Major oil discoveries in Texas, California, and overseas, in which Standard Oil had only minor interests, cut deep into the company's market share. The incandescent electric light bulb overtook kerosene lamps as the public's favorite lighting choice. Finally, with Rockefeller retired, the company lost its strategic advantage. Without their founder's leadership, they got a late start on the automobile craze highlighted by a slow transition from kerosene to gasoline. They were a major force, but no longer the dominant player.

The case reached the Supreme Court in 1911. Federal prosecutors had assembled a massive file of evidence backing the charges. Despite the media excitement that the federal government could nail one of the nation's most notorious robber barons, at the end of the day, the feds had almost nothing on the trust giant, particularly regarding the charges of predatory pricing and price gouging. The justices believed that Standard Oil's quest to drive down prices came from technology and their economies-of-scale advantage. As a result, manufacturing excellence and not Standard's predatory methods made profits difficult for its competitors. The hard truth was that Standard Oil did not raise prices once they eliminated or purchased the competition.

The Supreme Court struggled with how to punish a company committing the crime of providing Americans with an abundance of quality, yet affordable goods. The Court, presumably under public and political pressures, felt obligated to do something, so they determined that their volume discounts through rebates from the railroads 20 years earlier constituted a legitimate reason to penalize the large trust. SCOTUS punished The Standard's indiscretions by breaking up the company. By the way, it was illegal (per the ICC) for the railroads to furnish discounts, not for its customers to receive them.

The breakup included 33 *little Standards*, which were not so small and included Standard Oil of New York, renamed Mobil Oil, and Standard Oil of New Jersey, renamed Esso and later Exxon. Rockefeller retained significant stock ownership in all of them. After the breakup, the retired businessman earned a billion more dollars, becoming one of the richest men in history in time-adjusted currency. He lived to the age of 98, enough time to give most of it away. As noted earlier, even as a philanthropist, he was a businessman—the bulk of that charity funds he spent well.

The Standard Oil trial became a political and economic game changer. It instilled an anti-monopoly fervor into the American psyche—embedded into society's DNA. Apparently, that is what governments do—they protect its citizens from monster businesses that supply them with large volumes of quality and affordable goods. The first line of defense is to prevent monopolies from forming through mergers, but if the perpetrators slip through the cracks, then the DOJ goes on the offensive. Besides decimating the business, the government may fine the company's officers and even throw them in jail as felons. Strong terms such as predatory pricing, restraint of trade, and anti-competitive behavior reflect the seriousness of the offenses attributed to abusive businesses.

Despite the powerful rhetoric, antitrust or anti-monopoly theory is junk social science—pure fantasy. All industries consolidate as they move from craft production into automation. Previous chapters cited several examples, but there were many more. If Rockefeller never got into the oil business someone else would have consolidated the industry, even though the alternative probably would not have matched Standard Oil's high quality and low prices.

Antitrust legislation attacks the core of American exceptionalism. To anti-monopolists, providing a growing supply of affordable goods to the masses—the definition of economic growth—violates an unwritten mercantile-era code that discourages progress.

The concept behind antitrust theory has two stages. First, dominant businesses can charge cut-rate pricing (predatory) for the purpose of driving out competition. Second, with their rivals removed, they can charge higher prices (gouging) thus hurting consumers, while the perpetrators earn unnecessary, obscene profits. But is that how the system works?

Business leaders regularly commit numerous types of crimes involving commerce: bribery, breach of contract, fraud, blackmail, theft of intellectual property, tax evasion, extortion, and embezzlement to name a few. In those cases, the courts, not the market, must address real corruption. However, supply and demand, along with other economic laws, makes long-term price manipulation nearly impossible. In fact, the market system supplies consumers with all the tools necessary to keep big businesses in line.

Supply and demand laws dictate that the higher the volumes a business aims to achieve, then the lower the prices they must set. Economies-of-scale advantages provide the company with the potential to produce at lower prices. If a business raises their prices, even with reduced or little competition (there is always competition), those higher prices will drive away customers, thus reducing sales volumes and efficiencies, and therefore profits.

The Sherman Act, followed by the Clayton Antitrust Act (1914) and the Federal Trade Commission Act (1914), are silent prosperity killers. From 1911 on, since the Supreme Court ruled against Standard Oil, every business positioned at the pinnacle of their industry has to back off on their performance. The threat of massive fines, bad PR, and possible jail time takes away business leaders' incentives to pursue excellence. Later, as a segment of their "recovery strategy," New Deal administrators locked in on antitrust fervor as part of their mission to put economic progress in its place.

## Public Utility Companies

As the 1920s ended, politicians, bureaucrats, and social pundits turned their sights on the prolific electric utilities sector as a public-domain industry that should be owned and operated by municipalities. This local form of nationalizing industries swept through several large northeastern cities.

After World War I, rapidly growing electrical power companies overtook railroads as America's largest industry. Utility companies built the electrical grid to power light bulbs; however, a surge in home appliances and industrial machinery drove them to modify that strategy and move to supply significantly larger amounts of power. Meanwhile, new technologies provided power companies with the ability to deliver electricity more efficiently over longer distances and then seamlessly provide safer voltages to homes and businesses. The combination encouraged consolidation which involved mergers and acquisitions. Similar to the railroads, pricing often varied among geographic locations, time of year, suppliers, and customers.

Utilities fit the standard criteria for public regulation, or in the court's words, were "clothed in the public interest." For example, electric utilities served a large percentage of the public. They used well-established technologies, so the utilities and their suppliers had already done plenty of heavy lifting. For local governments it would be a "Thank you very much for your hard work, but now we'll take over" moment. Unlike regulation of the railroads, the move to regulate utilities proceeded rapidly and with potentially dire consequences as America's developing command economy had the potential to follow its European counterparts and stifle innovation.

Economist Harold Beirman Jr. (1918–2007) in his book *The Causes of the 1929 Stock Market Crash* (1998) notes that in the summer of 1929, utilities accounted for a considerable 47% of the value of the DJIA, and while the market was overpriced and vulnerable to a correction, utilities were more so. By the fall of that year, newspaper stories detailed the efforts of Boston and New York city selectmen to take over their city's private power utility companies. Beirman suggests that fears of government involvement, as much as any other factor, exaggerated the takedown of the mighty stock market.

The theory that regulation worries played a role in the partial collapse of the stock market is debatable, but the increasing interest by politicians to regulate private markets evolved into a real economic threat.

It had taken 50 years to build America's electrical power grid infrastructure, and yet, expensive work lay ahead. By the late 1920s, hundreds of private utilities provided electricity to most urban regions, and they were in the early stages of electrifying rural America. Even in metropolitan areas, utility companies' work was far from done as the appetite for electricity continued to grow. The solution to supply more power was to think big—provide cheap, abundant electricity by taking advantage of economies of scale. Big electrics' strategy of centralizing the electric-power industry made sense and it worked without regulation. Raising capital for the expensive infrastructure demanded a large market. The standard tool that business leaders employed was to consolidate smaller companies. Acquiring businesses that were already in existence saved money compared to the cost of running new cables next to a competitor's lines. For generating electricity, larger centralized plants cost less and were more efficient than many smaller ones

✪✪✪

Progressives had amassed the appropriate weapons, but to take their mission to the next level they needed an ally in the White House.

# 23

# Herbert Hoover the Wonder Boy

Few American presidential candidates have looked as good on paper as Herbert Hoover did in 1928. He was an extraordinary person, someone to admire. His resume came packed with the prerequisites that Americans sought in their nation's chief executive: a solid education and a productive career in the private sector. Later, Hoover became a well-known social organizer and humanitarian. His global travels placed him in the middle of historical events, which gave him extraordinary worldly experience. Hoover managed a prolific eight-year run as Secretary of Commerce. Plus, he was a successful author and dedicated family man. Yet, one event dominates Herbert Hoover's single presidential term—the Great Depression.

Hoover inherited a robust, surging economy, but on his watch, the stock market crashed. Then, along the economy's three-and-a-half-year downward spiral, the panic reached its lowest point. Where did Hoover go wrong?

Herbert H. Hoover was born in agricultural West Branch, Iowa, in 1874. His father, a blacksmith and farm equipment merchant, died young in 1880. His mother passed away four years later, leaving young Herbert an orphan at nine years old. Fortunately, an uncle of modest wealth living in Oregon took the boy in and raised him. In 1891, Hoover went off to college at the then brand-new Stanford University. The future President of the United States graduated in 1895 with a degree in geology.

After college, Hoover worked as a laborer at a California mine. The next year he landed a job as a geologist with a British mining firm. Within a few years, Hoover became a renowned mining engineer specializing in metals such as gold and zinc. His work took him to Australia, China, and England. While living in Tientsin, China in 1900, Hoover found himself, along with his wife Lou Hoover (1874–1944), at the center of the Boxer rebellion, a battle between Chinese nationalists and colonials. Caught in the urban fighting, the natural leader rallied his fellow Westerners to build barricades to protect their neighborhood until coalition forces came to rescue them.

Hoover's rising career path led to management positions. Later he became a part owner in various firms, an investor, and then a highly paid global consultant. The multifaceted Hoover even found time to write a popular textbook on geology, coauthored with his wife. The Great Engineer, as he became known, had made enough money that by 1914 he retired from mining. Financially secure, for his next venture the forty-year-old dedicated himself to humanitarian projects.

In his first mission, Herbert Hoover tackled evacuating Americans stranded in German-occupied Belgium during the early stages of World War I. His army of volunteers helped return home thousands of American citizens. The operation required volunteers and donations to help with food, clothing, passage, and cash.

Once the Americans arrived home, Hoover turned his sights to aid starving Europeans living in worn-torn Belgium and northern France. This international endeavor lasted two years and received funding from governments and private donations. Hoover's heroic effort included working long hours to make sure his organization fed nine-million people. As a leader in the charitable effort, he helped oversee the group's complex logistics. In addition, Hoover held numerous meetings in Germany with the occupation leaders to guarantee donated food went to the Belgian and French people and not to the German army.

Humanitarian work exhilarated Hoover. In 1917, he took that path to the next level as a high-profile public servant. That year, President Woodrow Wilson appointed the now-famous Hoover to lead the newly created United States Food Administration. The ravages of war led to food shortages overseas and potential shortages in America. Hoover's job was to make sure the country produced enough food for both civilians and the armed forces, with enough left over for Europeans. Farmers under Hoover's direction met all three goals. Particularly impressive were the 30,000+ tons of food provided to Europe during the war and reconstruction. Although he instituted funding to raise farm prices, he avoided food rationing by relying on his signature diplomatic tool, volunteerism. Hoover's volunteerism minimized government control and instead channeled the people's natural instincts to help one another.

As a private citizen, Hoover made another historic philanthropic gesture. In 1919, he donated $50,000 along with a treasure trove of European documents collected by the Allies during their European missions, to his alma mater Stanford University. Stanford used the funds and documents to establish the Hoover War Library, which later became the Hoover Institution, today the nation's leading conservative think tank.

Hoover, in his role as head of the Food Administration, worked for a Democratic president, so the Democrats pursued the progressive Hoover for their party. Instead, Hoover aligned himself with the Republican Party. He believed in private ownership as he loathed socialism and communism, but his progressive leanings led him to believe that the market needed oversight by the federal government.

When Warren Harding became President in 1921, he made Herbert Hoover Secretary of Commerce. In his new role, Hoover took a minor cabinet position and widened its scope. To accomplish his goals of a *more involved* government, he created sub agencies within his department, often overlapping the responsibilities of other cabinet departments. Under Hoover, the Department of Commerce promoted the exportation of American products. It assisted in negotiations during touchy labor disputes. Hoover worked with banking regulators and the construction industry to make home ownership more affordable. Hoover's agency also spent significant resources to work with American industry on the standardization of consumer products and regulating the airwaves. Each project followed the same pattern—a partnership between the private sector and government with a goal to reduce waste and make the economy stronger.

One of Hoover's more controversial initiatives as Commerce Secretary occurred in 1921 when he led an aid drive to feed the famine-stricken Soviet Union. Russians were starving to death under the Communist regime. Again, with the Soviets, Hoover augmented government funds with individual and corporate donations.

President Harding died in August 1923. Vice President Calvin Coolidge (1872–1933) became President and kept Hoover on as Secretary of Commerce for another 6 years.

In 1927, a massive flood, the worst in American history, devastated the lower Mississippi River basin. Secretary Hoover sprang into action and helped lead an effort to supply food, shelter, and clothing for hundreds of thousands of displaced Americans. For his role in the flood relief, Hoover made the March 26, 1928, cover of *Time Magazine*.

Coolidge, with his hands-off approach to governing, did not agree with Hoover's brand of political thinking. In a derogatory tone, the President often referred to Hoover as Wonder Boy. Outside of the oval office Hoover had a similar reputation. Among his colleagues, he earned the title of *Secretary of Commerce and Under Secretary of Everything Else*.

Despite the backlash to his aggressive style, Hoover continued to gain political capital. In 1928, he won the Republican nomination for President. Hoover ran on the platform that he could make a strong America even stronger. Buoyed by that robust economy, Hoover won the presidential election in November. He dominated both the popular vote and the Electoral College.

✪✪✪

On a cool, drizzly afternoon on March 4, 1929, after being sworn in by Chief

Justice, and former president Howard Taft (1857–1930), President Herbert Hoover stood by, prepared to deliver his inaugural address. Standing on the East Portico of the US Capital, Hoover opened his speech to the large, rain-soaked crowd by praising the American economy, the American people, and the American government. He gave a brief nod of appreciation to his predecessor Calvin Coolidge for his "wise guidance." Fifteen minutes later, President Hoover ended his keynote speech in a similar tone, "I have no fears for the future of our country. It is bright with hope."

While acknowledging his inheritance of a strong economy during the introduction and closing, Hoover's main thesis, contained in the body of his March 4th address, claimed a more somber mood—oddly pessimistic. He spent an excessive amount of time focused on how it was time for the federal government to take the lead on matters that the previous administration believed should have been the responsibility of either local governments or the private sector. Hoover suggested that the federal government get more involved in education, health coverage, welfare, policing crime, and farm subsidies. Though Hoover warned that government should not own industry, he declared three times in his speech that government must increase its regulation of the private sector. Most revealing is how much the President displayed a lack of confidence for the individualistic nature of the American people. For example, he warned that "self-government can succeed only through an instructed electorate." The core message from his inaugural address may have been that Hoover thought a great deal of himself, and not so much of the extraordinary American economic system or the citizens who elected him.

# 24
# Smoot-Hawley Launches the Great Depression

Following the economy's brief recovery in May 1930, stock values again plunged, shadowed by other economic indicators. Banks foreclosed on thousands of farms while businesses throughout all industries struggled. The economic downturn inflicted financial havoc as banks closed and deflation devalued the dollar by one third. Unemployment, which had hovered around 4% for most of the 1920s, reached 15% by year's end. The economic depression deepened in both 1931 and 1932. The estimated unemployment rate over that time exceeded 20%.

Based on history, two years should have been plenty of time for the free market to turn the economy around, but 1920s-type prosperity would not return until after the war. Escalating an *ordinary* recession into a lengthy, severe panic required a series of extraordinary political blunders that stretched well beyond ill-managed government. To impede the most robust economy in world history required the nation's leaders to apply a blatant disregard for common sense and to neglect intellectual due diligence. The inaugural blunder of all those Great Depression political sins occurred on June 17, 1930, the day President Hoover signed into law the Republican-backed Smoot-Hawley Tariff Act (also referred to as the Hawley-Smoot Tariff Act).

Congress constructed Smoot-Hawley as a protectionist trade initiative, which placed a significant import duty (tax) on foreign goods that were in direct competition against domestic businesses. The bill's sponsors, Senator Reed Smoot (1862–1941, R Utah), and Congressman Willis C. Hawley (1864–1941, R Oregon), believed that with diminished competition, ailing agricultural and manufacturing sectors could charge higher prices and reverse the trend of deflation and sluggish domestic sales while providing higher wages for workers. Hoover supported Smoot-Hawley and believed increased import duties would slow down the drain of the treasury's gold reserves, another problem caused by deflation.

The controversies over Smoot-Hawley's excessive duties produced spectacular debates—was it good or bad legislation? Adam Smith, the father of modern economics, weighed in on the controversy 154 years earlier.

In *The Wealth of Nations* Smith writes,

> In every country, it always is and must be the interest of the great body of the people to buy whatever they want of those who sell it cheapest. The problem is so very manifest, that it seems ridiculous to take any pains to prove it...

The United States Congress, throughout its history, included both protectionist and free-trade proponents. In the early days of the republic, the federal government used tariffs as a source of income and a tool to protect and nurture infant industries. Interest in protectionist tariffs increased with mid-nineteenth century industrialization. The primary motivation came from Northern Republicans. After 1913, the Sixteenth Amendment (income tax) reduced the need for government revenue from tariffs, but the concept that the heavy taxation of imports boosts economic growth retained modest popularity.

By 1930, most scholastic economists disapproved of protectionism in favor of free trade. Other scholars, such as historians and archaeologists, likewise shared an appreciation for unabated international trade. On the political side, Southern Democrats, along with members of both parties from the heavily agricultural states opposed tariffs.

Those people understood that open international trade was more than just an economic benefit, it was crucial for national prosperity. Their free-trade viewpoint emphasized that fewer trade restrictions resulted in greater economic benefits for both trading nations.

Did trade commentaries contribute to the severity of the October stock market crash? The *New York Times* started editorial chatter about a forthcoming aggressive tariff bill in January 1929. During the tariff bill's Congressional debates on May 5, 1930, the *New York Times* published a front-page story entitled, "1,028 Economists Ask Hoover to Veto Pending Tariff Bill," written by a group of free traders fiercely opposed to Smoot-Hawley. In the article, a who's who of American economists warned the public that the protectionist bill "would therefore raise the cost of living and injure the majority of our citizens...The vast majority of farmers, also, would lose...Our export trade, in general, would suffer." The group, which included professors representing 179 colleges and leaders from 46 states, also cautioned that higher American tariffs would "bring reprisals."

Hoover got an earful from his various confidants as well. Presidential advisor Thomas Lamont (1870–1948), of the investment firm J.P. Morgan, begged Hoover to veto the "asinine Hawley-Smoot Tariff." Industrialist Henry Ford told the President the bill was "an economic stupidity." Foreign governments weighed in with their opposition. Despite the wise advice, Hoover remained steadfast in the protectionist camp.

As with past trade bills, bipartisan protectionists, or as they now prefer to call the issue, *managed trade* in an attempt to distance themselves from protectionism, reacted not to sound social science, but to high-profile business failures and industry-wide job losses. Influence came not from intellectual thought, but through lobbyists and special interest groups. In the end, political favors and other types of corruption set trade policy.

Smoot-Hawley became so all-encompassing because its architects designed their tariff legislation to protect not only factories but farmers—the latter of whom had minimal foreign competition. With *something for everyone,* the high-tariff advocates enticed enough votes from Southerners and Mid-westerners to pass the legislation.

The infamous trade bill applied those high tariffs on nearly 3,300 items, one-third of America's total imports. At its peak in 1932, the average tax on those dutiable goods approached 60%, or about 20% across all imports.

A punitive tax method, known as *ad-valorem,* proved especially harmful to shrinking imports and exports. Smoot-Hawley levied a percentage rate against some imports, but on others it charged a flat rate—*ad-valorem.* In the latter case, as the rate of deflation increased through the early 1930s, so did the tax rate. See Table 24.1 for an example of how *ad-valorem* tariffs increase during deflationary periods.

**Table 24.1** Deflationary effects of *ad valorem* tariff

| Year | Price | Tax | Percent |
|------|-------|-----|---------|
| 1930 | $1.00 | $0.25 | 25% |
| 1932 | $0.50 | $0.25 | 50% |

Smoot-Hawley did not start out as an economic disaster. In fact, early on its proponents believed their plan had worked. The economic slide seemed to have leveled off—not so much as a burst of recovery, but there were subtle signals of hope. Nevertheless, higher prices did not help. Instead, the failure rate of farms and factories increased. Wages did not increase either. They fell while the unemployment rate continued to climb. In desperation, federal politicians, still the protectionism optimists, further increased import duties. Once again, the higher tariffs sank the economy deeper into economic depression.

The launch of Smoot-Hawley caused more harm than just high tax rates and inflated prices. The hefty duties imposed on America's allied nations' exports outraged America's trading partners, and, as predicted; they unleashed their *reprisals.* Nations, most notably Canada, levied their own retaliatory tariffs on American imports. In response, to offset a shortfall in American imports and exports, countries formed their own trading blocs with *favored-nation* status that excluded the United States. Those retaliatory actions not only damaged the American economy, but they triggered a devastating chain reaction that caused both imports and exports to drop around the world, swelling the global economic depression.

Today, unlike Roosevelt's still-popular New Deal initiatives, Smoot-Hawley keeps its dreadful reputation throughout the academic world as an aggravating contributor to the Great Depression. Although despite the majority view, few historians and other scholars maintain that Smoot-Hawley was a major cause of the economy's downfall. Overall, conventional wisdom downplays Smoot-Hawley—as a *mere nuisance* for its role in the Great Depression. The small percentage of post-Great Depression academics that consider the tariff act a core contributor to the economy's demise looks to be out of proportion compared to Smoot-Hawley's negative reputation. In fact, two of the most noteworthy challenges to scholarship's subdued view come from outside the academic world.

One came during a presidential debate, delivered as a knockout punch by Vice President Al Gore (b. 1948) to Ross Perot (1930-2019), the Independent Party candidate. On November 9, 1993, Gore and Perot debated the proposed North American Free Trade Agreement (NAFTA) on CNN's *Larry King Live*. The current VP accused his adversary of being a protectionist, citing the Smoot-Hawley Tariff Act as evidence of the dangers when pursuing that objective. For dramatic effect, Al Gore presented Perot with a framed picture of Mr. Smoot and Mr. Hawley. The Vice President noted how the two men, "sounded reasonable at the time." Gore then claimed that Smoot-Hawley "was one of the principal causes, many economists say, the principal cause, of the Great Depression…"

Another famous analysis blaming the 1930 debacle for the Great Depression comes from Hollywood in the 1986 blockbuster movie *Ferris Bueller's Day Off*. In a legendary scene, a high school economics teacher, played by real-life political pundit and comedian Ben Stein (b. 1944), attempts to educate a classroom full of disinterested students. The actor presents the lesson in his trademarked deadpan voice.

> In 1930, the Republican-controlled House of Representatives, in an effort to alleviate the effects of the… Anyone? Anyone?… the Great Depression, passed the… Anyone? Anyone? The tariff bill? The Hawley-Smoot Tariff Act? Which, anyone? Raised or lowered?… raised tariffs, in an effort to collect more

revenue for the federal government. Did it work? Anyone? Anyone know the effects? It did not work, and the United States sank deeper into the Great Depression.

Various academic fields have rationalized why Smoot-Hawley's massive taxes on imports does not rank as a major cause of the Great Depression. But do their reasons hold up to scrutiny?

Historians discount Smoot-Hawley's damage based on what they believe to be more noteworthy catalysts that preceded the 1930 tariff act. The culprits they focus on are capitalism's *recent failures*. The long list includes: the farm crisis, the stock market crash, bank closures, fluctuating gold reserves, and a decline in the dollar's value. The truth is, capitalism was thriving prior to the Great Depression, there was no farm crisis, and stocks were turning toward a bull market by mid-year 1930. Later, the New Deal addressed bank failures, the gold standard, and deflationary monetary policies with little success. Therefore, amongst all the financial matters, all were resolved by 1934, long before the economy recovered. Whereas the recovery occurred during a time of low tariffs.

Mainstream economists use simple arithmetic to temper Smoot-Hawley. They argue that in 1930 foreign exports accounted for a mere 5% of the American economy—an amount much less than the Great Depression's 40% shortfall. This presumption's major flaw is that the simple arithmetic does not explain trade's ripple effect. Consider a modern-day hypothetical situation. Petroleum production accounts for about 5% of the early twenty-first-century US economy. By the same math, who would argue that a 40% drop in oil production would lead to only a 2% slump (economist's math) in the American economy? In reality, a single-digit dip in petroleum products would cause a severe economic depression.

Populist free traders downplay Smoot-Hawley by citing that Congress had passed high-tariff bills in the past with no ensuing depressions. That is true, but there were several noticeable differences separating Smoot-Hawley from previous high-tariff protectionist legislation. First, there was its bad timing. Congress passed Smoot-Hawley on the downside of an economic bubble. Second, earlier tariff bills approaching the scale of Smoot-Hawley occurred a century earlier at a time when America's less-industrialized economy was not so dependent on foreign trade. Third, severe deflation caused a dramatic tax increase on the many products protected by ad-valorem tariffs. Finally, the extreme retaliation to Smoot-Hawley by America's trading partners caused a dramatic drop in both imports and exports.

There is another possibility for conventional wisdom's lack of conviction toward the 1930s tariff act as a major cause of the Great Depression, albeit this one has a sizable conspiracy theory component. If Smoot-Hawley had provoked the rapid economic decline that began after May 1930, then the early depression's antidote would be its repeal—a viable alternative to the New Deal. In fact, FDR, during his 1932 presidential campaign, threatened to overturn the Republicans' Smoot-Hawley Act. Once in office the President abandoned his protectionist challenge in favor of a more liberal agenda. Academics did the same.

# 25

# FDR Moves into the White House

On March 4, 1933, Franklin Roosevelt succeeded the one-term Herbert Hoover to become the 32nd President of the United States. His revolutionary presidency became the longest in American history. To his credit, historians consistently rank FDR as the third greatest president, trailing national icons George Washington and Abraham Lincoln.

FDR took office with the nation mired deep in the Great Depression. His New Deal so radicalized the role of the federal government that much of the recovery model continues to drive national economic policy. He then took on a second legendary chapter as a wartime commander-in-chief. In that latter role, he led the armed forces and reassured the nation during America's most intense war with foreign nations. America's chief executive met a sudden death just one month before the German surrender and six months prior to the dropping of the Hiroshima and Nagasaki atomic bombs.

Franklin Delano Roosevelt was born to a wealthy family in Hyde Park, New York, in 1882. Their surname came from seventeenth-century Dutch-American ancestry. President Theodore "Teddy" Roosevelt (1858–1919) was a fifth cousin. Thanks to his family's affluence, Roosevelt received an excellent private education. His interests at school ranged from foreign languages to sports. Given his extraordinary opportunities, historians consider his school year achievements mediocre. Financial privilege also gave the young man some worldly experience through annual family trips to Europe. FDR attended Harvard University, graduating with a degree in history in 1903. He then studied law at Columbia College. In 1907, Roosevelt passed the New York State Bar exam. After college, he dabbled in his craft at a law firm where he worked at an entry-level position.

In 1905, Roosevelt married a distant cousin and the then-serving President Teddy Roosevelt's favorite niece, Anna Eleanor Roosevelt (1884–1962). Eleanor would later become a powerful figure in the Democratic Party. The couple had six children; a boy Franklin died at six months old.

Following his uneventful stint at the law firm, Roosevelt decided that his heart was in politics. In 1910, Roosevelt, running as a Democrat, won his first try at a public office, a seat as a New York state senator. His relation to President Roosevelt no doubt helped his victory. Two years later, the rising star, while campaigning for Woodrow Wilson, caught the next President's attention. The following year, Wilson appointed Roosevelt Assistant Secretary of the Navy, a post he held through World War I and up to 1920. In Hoover-like fashion, FDR served as an effective and forward-thinking naval administrator. He became an expert in US Navy history and initiated numerous positive reforms.

Roosevelt again entered politics in 1920, as James Cox's running mate. Cox lost the presidential bid to the Republican Warren G. Harding. FDR considered this heartbreaking vice-presidential loss a serious setback. Then, less than a year later, a real tragedy struck. On a Canadian vacation, the future President contracted polio. He was 39 years old. The misfortune put a temporary halt to FDR's ten-year foray as a politician and bureaucrat.

The infection left Roosevelt paralyzed from the waist down for the rest of his life. Determined to hide the disability, Roosevelt would wear heavy leg braces so he could stand and even imitate a walk with the help of a cane in one hand and the support of an aid in the other. Even though confined to a wheelchair, few photographs exist showing him sitting in one.

For the next several years, FDR refrained from seeking public office, yet he remained active in politics, promoting candidates for his Democratic Party. During this period, he also tried several business ventures. All of them failed. In 1928, Roosevelt returned to the political fray. In his first contest, the just-left-of-center progressive ran for governor of New York and won. FDR turned his new high-profile position into a stepping-stone toward the presidency. To that aim, the savvy Roosevelt built a formidable coalition of political allies. He would later use their influence during his run to the 1932 Democratic Presidential nomination.

The 1932 Democratic candidate rallied on a pledge promising to pull the country out of its current deep recession, which he blamed on President Hoover. Ironically, FDR campaigned against Hoover on the platform declaring that he would rein in excessive, wasteful federal spending. Roosevelt characterized the Republican administration as having too many extravagant federal agencies funded by a reckless fiscal policy. Just two months before the election, Roosevelt attacked Hoover's spending policy by proclaiming, "I accuse the recent Administration of being the greatest spending Administration in peacetime in all American history." That next month, the progressive contender reiterated his opinion: "I regard reduction in Federal spending as one of the most important issues in this campaign." FDR was spot on. With a focus on reviving the economy, Hoover had launched a series of expensive infrastructure projects, such as the Boulder Dam (renamed the Hoover Dam in 1947). With the Republicans in charge, federal government spending rose a hefty 30% during the 1932 campaign year.

Later, Roosevelt switched gears, moved left, and focused on a series of progressive initiatives designed to fix the broken economy. He even made the bold promise to implement those reforms within his first 100 days in office, which became the out-of-the-gate standard for future presidents. FDR told the citizens that if elected, he would reinvent government with what he would later coin as a *New Deal*. His goal, in the candidate's words, was "to reform, regulate, and restore," the ailing economy.

Just as Hoover couldn't lose in 1928 thanks to a stellar economic outlook, he couldn't win in 1932 for the opposite reasons—the now dismal economic conditions of the three-year recession. The tell-all statistics for the fall of 1932 were the unemployment rate which exceeded an estimated staggering 25%, and a plummeting stock market, which had dropped 90% from its peak in August 1928. The poor state of the economy, combined with FDR's astute political posturing and superb speaking skills, won the former Democratic New York governor the presidency in a 472 to 59 electoral landslide.

Roosevelt, unlike Hoover, was not the deep-thinking philosopher; however, he was a much more effective national leader. FDR stood 6 feet 2 inches tall with a strong upper body and a firm handshake, the latter, in part, to compensate for his disability. Many noted Roosevelt had a broad and warm smile. FDR had a politician's deep voice, which was amplified by Roosevelt's lifelong addiction as a chain smoker. The President worked on building a sincere rapport with the people as a man they could trust and someone capable of doing something about the poor state of the economy. In his first inaugural address, the President demonstrated those traits, "let me assert my firm belief that the only thing we have to fear is fear itself..." His regular fireside chats became legendary. FDR made hundreds of radio broadcasts to first assure the American people that the health of the economy was improving and then later as a vehicle to comfort a nervous nation at war.

While his pedigree helped with many of FDR's early opportunities, he turned into a powerful force on his own. As president, FDR's approach to leadership became his most notable legacy. Before the war, Roosevelt sought two goals: fix the economy and advance the Democratic Party's agenda. FDR maintained his focus for the rest of the decade to achieve those goals. He used two tools to accomplish those missions.

To achieve the first goal, FDR built his support staff, a troupe of advisors and consultants to develop his battle plans. He called the posse his brain trust. Half of them had a Harvard College education, and most were lawyers, professors, or both. The group included only one woman, Frances Perkins (1880–1965), although that was not unusual for the era. Not the wide range of diversity one would expect in a group responsible for redefining a heavily industrialized economy. The brain trust then employed an army of administrators whom he set to work building a massive central bureaucracy.

Roosevelt's second approach was to do whatever it took to move his agendas forward. FDR would sometimes bend America's rule of law and even circumvent the United States Constitution. He infamously tried to pack the Supreme Court with liberal judges. He used diplomats and advisors as pawns to *test the waters*. Whether those actions were justified often depended on one's political viewpoint. To maintain his political momentum, FDR countered tradition to serve not just a second, but a third, and then began a fourth presidential term.

Of his two goals, the economy and the liberal agenda, Roosevelt had split success. On his first objective, full economic recovery, the President did not fare well. The unemployment rate fell following his first inauguration; however, over the next 7 years, from 1933 through 1939, the unemployment rate averaged 18% and never fell below 14%. For the second objective, his progressive agenda, FDR had great success. Despite some New Deal setbacks by the Supreme Court, the Republicans, and conservative Democrats, the country moved noticeably left, for which FDR can take full credit.

Roosevelt set the tone for his accomplished legacy early on. New laws and executive orders came fast and furious. In a cooperative frenzy covering its first 100 days, the FDR administration and Congress passed 15 significant laws that:

- Ended prohibition.
- Shut down the banking system and revamped it, which included the Federal Deposit Insurance Corporation. (FDIC)
- Regulated Wall Street.

They established the well-known alphabet agencies including:

- The Federal Emergency Relief Administration (FERA) that helped support soup kitchens.
- The Emergency Conservation Work (ECW), renamed the Civilian Conservation Corps (CCC) in 1937, that put millions of the unemployed back to work building roads, planting trees, and a host of other jobs deemed valuable to communities.

- The Agricultural Adjustment Administration (AAA) who directed large sums of money toward farming to reverse the falling prices of agricultural-goods.
- The National Industrial Recovery Act (NIRA) that funded infrastructure projects and attempted to reform industries.
- The Tennessee Valley Authority (TVA) who provided flood control, electricity, and economic growth to one of America's poorest regions.

Despite its economic failures, the New Deal was influential enough to create a noticeable shift in the nation's political ideology—the great political paradox—a certified disaster turned into a resounding hit. To add more intrigue to the political narrative, consider that FDR entered the presidential contest with centrist leanings. The President's journey from a moderate progressive governor of a Republican state to an ideologue on the edge of socialism had an interesting twist.

As noted, Franklin D. Roosevelt, began his first presidential campaign with a conservative tone, chastising Hoover for reckless spending and over-regulation. Once in office, FDR again showcased his moderate-progressive leanings with his tax policy. True, the President considered excessive, heavy taxes on the rich justifiable, per the customary position of the Democratic Party, but he also understood the lower-income public should pay their fair share.

High taxation was an integral part of progressive ideology. Roosevelt inherited an aggressive federal tax code from the Hoover administration—he then added to it. He tripled the capital gains tax, which included significant increases on minimum exemptions. FDR established the undistributed corporate profits tax, aimed at diverting funds allocated for future private investment and funneling them into the New Deal.

Not only punishing the rich, federal taxes hit the average citizen hard. A slew of new excise taxes became a major source of New Deal funding. New-found taxed items included food, drinks, cigarettes, tires, cars, electricity, and phone calls. Each of them burdened the poor more than the rich. The new unemployment tax and social security tax also tapped a higher percentage of the working man's paycheck. While they went the route of *politically convenient* hidden taxes for which the employer *paid* the entire unemployment tax and a half of the social security tax, both popular programs had tax rates that were regressive (the poor paid a higher percentage than the rich).

Social Security exemplifies FDR's commitment to his everybody-should-pay-taxes policy. His golden, third-leg-of-a-three-legged-stool program not only featured a regressive tax rate, but the tax began in 1937, while the government did not issue the first benefit checks until 1940.

The liberal concept of high taxes devastated the 1930s economy but holding citizens responsible for government spending was a conservative ideal. It was not until America's involvement in World War II that the treasury adopted Keynesian easy-money policies, which tax-shy politicians ramped up after 1971 hiding an ever-increasing portion of the tax burden with deficit spending.

Two more examples of FDR's conservative leanings are that he initially opposed the FDIC, considering it a bailout for the banking industry, and throughout his presidencies, Roosevelt opposed labor unions for government workers. But the inspiration for the New Deal is yet another example of interesting, Great Depression history.

As it turns out, the New Deal started its crusade during Herbert Hoover's *do-nothing* presidency. During his stints as Secretary of Commerce and US President, Hoover introduced many of the programs that became the New Deal. New Deal architect Rex Tugwell (1891–1979) and fellow brain trust academic Raymond Moley (1886–1975) made the following notable statements crediting Hoover with many of the New Deal inspirations.

Moley wrote in a *Newsweek Magazine* article dated June 14, 1948, "When we all burst into Washington… we found every essential idea (of the New Deal) enacted in the 100-day Congress in the Hoover administration itself. The essentials of the NRA (National Recovery Administration), the PWA (Public Works Administration), and the emergency relief setup were all there. Even the AAA (Agricultural Adjustment Act) was known to the US Department of Agriculture. Only the TVA and the Securities Act were drawn from other sources. The RFC (Reconstruction Finance Corporation), probably the greatest recovery agency, was of course a Hoover measure, passed long before the inauguration." Economist Steven Horwitz (b. 1964) in his paper *Herbert Hoover, Father of the New Deal* references a letter written by Tugwell to Moley in 1965: "we were too hard on a man who really invented most of the devices we used."

Besides Hoover, where did the driving force for the New Deal come from? As the first few chapters of this section reveal, progressive ideologies had been building momentum for some time. The people behind the movement might be described best as technocrats. *Merriam-Webster* defines a technocrat as "a technical expert; especially: one exercising managerial authority." The term often applies to left-wing ideologues made up of politicians, public-policy scholars, and pundits. Their specialty—make nonsense appear plausible.

When the insurgence of technocrats reached a critical mass, they formed an *expertocracy*, a group powerful enough to become both the architects and the critics of the New Deal.

✣✣✣

The New Deal took on a tough challenge—suppress an economy with enormous potential and hungry for a rebound. To that aim, the next six chapters review some of those initiatives that helped extend the Great Depression. Note that despite their failures, the expertocracy deemed much of the social experiment worth keeping, and with the help of their authority, New Deal programs persevered—continuing to muffle twenty-first century economic growth.

# 26
# Financial Follies

FDR's brain trust blamed America's financial institutions for escalating the chaotic economic conditions that launched the Great Depression. Modern experts agreed. According to them, the primary culprits were the Federal Reserve Board, private banks, and the national gold reserves. The plea to bolster the US dollar made financial reform a priority and therefore an integral part of the President's renowned *First 100 Days* New Deal package. Today, academics consider those progressive reforms vital to the 1930s economic recovery and necessary for modern good economic health. However, two benchmarks, the severity of the 1937-8 recession (Table 26.1) and the continuing high unemployment rate throughout the 1930s (Table 26.2), suggest their reforms produced lackluster results. Did the New Dealers have the justification to restructure American financial institutions?

**Table 26.1** A brief comparison of the 1929 and 1937 recessions

|  | 1929 Recession | 1937 Recession |
|---|---|---|
| DJIA High | 381.2 | 190 |
| DJIA Low | 198.7 | 99 |
| % Unemployed Low | 3.2 | 14.5 |
| % Unemployed High | 8.7 | 19 |

**Table 26.2** Great Depression unemployment rate highlights

|  | 1933 | 1934 | 1939 |
|---|---|---|---|
| % Unemployed | 24.9 | 21.7 | 17.2 |

## The Fed

According to Great Depression scholars, at its best the Fed did little to prevent the 1929 economic crash, but at its worst, America's central bank loosened the money supply in mid-1929, thereby boosting the surge, and then tightened the money supply in late-1929, preventing a speedy recovery. If true, they practiced a pro-inflationary policy during a period of high inflation and a pro-deflationary policy during a period of high deflation. What is the Fed and what could it have done?

The Constitution, in Article I, Section 8, authorized Congress, "To coin Money..." and to "regulate the value thereof..." To help accomplish those goals, Alexander Hamilton, America's first Secretary of the Treasury, championed the formation of the first national bank. Over the next 125 years, the

Congressional and Executive branches established and closed a few national banking systems, although they were all weak versions of a true central bank.

America, from the beginning, had established an extensive network of sound private banks, so that in practice, a central bank would support the private sector for emergencies with influxes of currency as needed. On the other hand, a poorly designed central banking system could prove ineffective yet supply private banks with excessive profits.

After a series of severe economic recessions in the late nineteenth and early twentieth centuries, capped by the Panic of 1907, influential members of the financial community applied greater pressure on Congress to establish a prominent central bank. They envisioned the central bank as both a regulator and as a national depository for surplus currency—the latter to be used to smooth the economy's peaks and valleys. Congress established the Federal Reserve System, also known as the Fed, through the Federal Reserve Act of 1913. The Fed became operational in late 1914.

The Federal Reserve could:

- Regulate private banks
- Standardize the fractional reserve rate
- Lend money to banks

In theory, the Fed had the authority and the means to dampen recessions by raising interest rates during a surge and lowering them during the collapse, and with the latter, adding cash to the economy. Its stakeholders also believed the Fed could lower unemployment rates.

Of course, the Feds failed in 1929, but there should have been lessons learned and applied to the 1937 crash. The subsequent recession of 1937-38 and continuing high unemployment proved they were still not up to the task of improving economic wellbeing.

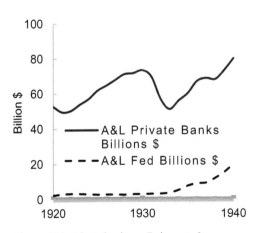

**Figure 26.1** Private banks vs. Fed assets & liabilities, 1920-1940

Figure 26.1 compares the assets and liabilities of private banks to those held by the Fed. The data covering the Great Depression's early years indicates that the private banks held the vast majority of monetary power and that the meager funds allotted to the Fed were no match for the mighty 1929 recession. However, notice the uptick in the Fed's assets during the New Deal years—sort of puzzling. By 1937, the Fed had tripled its funding and then doubled that 1937 amount by 1940, yet it could not stabilize the economy. The beefed-up Fed's lack of results raises questions—did the Fed need more clout or was any central bank incapable of promoting ideal Keynesian economic health?

## The Banking Crisis

To mainstream academia, the fact that many banks failed during the first four years of the Great Depression, from late 1929 through 1933, gives them substantial evidence of a failed, archaic institution in the free market system unable to support economic prosperity. The numbers are staggering. According to the US Census Bureau, approximately 8,000 banks suspended operations during that time, half of them in 1933. In total, bank suspensions during the first four years of the Great Depression accounted for about 30% of the nation's banks. One important note—the census data does not specify the nature of the bank suspensions. Some banks reopened after a cooling-off period or earned a second life through mergers. Others shut down permanently, sometimes resulting in customers losing their savings.

Either way, the fallout from the banking disaster has been well publicized as a major contributor to the ensuing economic turmoil. After all, the banking crisis contributed to the financial ruin of countless families. Then, as the banks' ability to loan money diminished, the nation forfeited some of its capacity to promote economic growth. Subsequently, public panic set in and bank customers withdrew their money, causing even more banks to fail. In an escalating pattern, America's money supply further contracted, fueling the ongoing deflationary trend. Once again there is more to the story in this case—the string of bank failures predates the Great Depression.

Though it is true that thousands of banks closed during the early years of the Great Depression, another 6,000 banks closed between 1923 through 1929 (see Figure 26.2). Great Depression academic enthusiasts know these statistics well, but general historians often gloss over them. In 1926 alone, a year of low

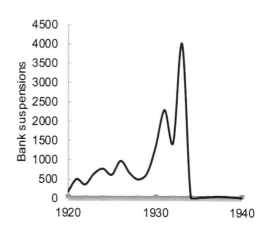

**Figure 26.2** Number of bank suspensions per year, 1920-1940

unemployment and high economic growth, nearly 1,000 banks shut their doors. One of the primary reasons is even more interesting.

There should be minimal doubt that antiquated state regulations contributed to the banking crisis. At the time, many states outlawed branch banking, the banking system we have today. In the 1920s and early 1930s, state regulators preferred unit banking, which is a banking enterprise of one bank. Regulators believed the policy best served America's small communities. The problem is that diversity is the lifeblood of banking. Prohibiting branch banking placed the wellbeing of a unit bank at the mercy of a single local economy.

Economist Jim Powell (b. 1944), in his book *FDR's Folly: How Roosevelt and His New Deal Prolonged the Great Depression* (2003), notes that although Canada suffered an equivalent economic panic to that of the United States during the same period, no banks closed their doors. In 1930, Canada had eleven main banks which controlled the nation's 4,040 branches. Powell also noted that in America, banks in states that allowed branch banks had lower closure rates.

That said, 1933 was a horrific year for banks as 4,000 suspended some level of operations. It was also the token year for high unemployment. Did bank suspensions deepen the depression, or did the depression increase the rate of bank suspensions? The answer is not clear.

In response to the banking crisis, the Roosevelt administration directed a series of sweeping changes designed to revamp America's monetary systems. The President signed an executive order mandating banking holidays to give federal agencies time to assess the problem. Later, Congress passed the Banking Act of 1933, which gave the Federal Reserve the authority to enact deposit insurance, which guaranteed the security of most saving accounts.

By the mid-1930s, America took a cue from the rest of the Western world and shifted toward branch banking. Still, the data in Figure 26.2 contradicts the Great Depression storyline. The number of bank suspensions plummeted, while the economic panic lingered on.

## The Gold Crisis

To the practical economist, gold serves as a near-perfect form of currency. The precious metal has enough commonality for widespread distribution with the durability to remain useful for centuries. However, its most redeeming characteristic, as a reliable medium of exchange, is the heavy metal's scarcity. There are enough golden nuggets scattered about the earth to keep strong economies moving along at a steady pace, but not so much as to ignite extended bouts of inflation. Silver has similar qualities and often serves to augment gold reserves.

Politicians and bureaucrats have not always been fans of gold's attributes.

For ages, mostly to fund wars but also to push social agendas, governments have weakened those qualities in favor of more *available* funds—either by diluting gold with common metals (debasing), or via the printing press and ledger (fiat currency). When properly executed, politicians can pad their government's pockets with extra revenue without having to levy additional taxation on their citizens. In Europe, the massive expenses of World War I created one of those scenarios.

In 1914, the world's developed nations adopted a fixed gold standard, with London serving as the financial capital. Each nation tied the value of their currencies to the British standard of an ounce of gold. The benchmark did not last long. In a change of heart, Great Britain, adopted emergency measures to fund World War I debt, and dropped the gold standard to pay for the war effort. After the war, and despite their earlier discrepancies, England wanted to return to the gold standard at the 1914 price. England, though, had created too much money, making a straight conversion nearly impossible. Other European nations faced similar problems. Meanwhile, America stayed on the 1914 Gold Standard after the war. Because the US had significant gold reserves and a stable currency, the financial center of the world shifted toward New York City. Uncomfortable with the new arrangement, the United States central bank used its financial power to help Great Britain and other European nations lock in their currency at pre-war prices. The helping hand did not go as planned. The gold dilemma caused excessive fluctuations in gold reserves as the precious metal moved back and forth across the ocean. A common criticism of 1920s US financial policy is that those fluctuations helped cause the Great Depression, or at least aggravated the stock market crash.

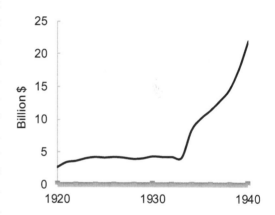

**Figure 26.3** The value of gold held by the United States, 1920-1940

FDR's brain trust agreed but they also considered a strict gold standard problematic to a well-run economy. Their response to the gold *crisis*, was to increase the price of gold from the established amount of $24.17 an ounce to $35 an ounce. Figure 26.3 illustrates that the trend continued throughout the 1930s.

The value of America's gold stock soared throughout the late 1930s—and once again with lackluster results.

## Conclusion

Those in favor of financial reform got everything they wanted—easy money and regulation; yet, the Great Depression stuck around. Eight years later, with all the reforms well entrenched, America entered World War II still stuck in an economic slump.

# 27
# Labor's Friend

New Dealers considered the labor unions' fight to organize workers a priority. They focused on federal legislation that gave labor unions leverage against powerful employers. But Congress had to be careful. They also needed to stop the cycle of violence against society by militant strikers.

Not only did constituents task Washington with solving double-digit unemployment, politicians had to negotiate their way through the politically charged issue of labor unions. While the New Deal did not fare well solving the unemployment problem, it had a resounding success promoting organized labor and stemming the hostility that accompanied it. No small feat—the New Deal revolutionized American labor. It had been a long battle.

Sixty years earlier, the Gilded Age with its mega-sized industries spawned a new chapter in the American labor movement. During that period, the ongoing battles between low-income workers and those with greater privilege became widespread and violent. To gain perspective into workers' crusades, the following pages summarize famous labor events of the later nineteenth and early twentieth centuries. Although these examples represent only a small sampling of the tens of thousands of labor disturbances that occurred between 1870 and 1934, they add historical perspective to the labor movement.

### 1877–Great Railroad Strike of 1877–Maryland, West Virginia, Missouri, and Illinois

The Great Strike became one of America's earliest, large-scale, and most notorious labor disputes. At the time, a small percentage of workers, including railroad workers, belonged to a labor union. As a result, the leaderless disturbance resembled a non-franchised collection of violent riots rather than a well-organized strike.

Leading up to the Great Strike, workers tried to reverse a 10% pay cut, the second one initiated by the railroads in response to the deflationary pressures of the 1873 recession, America's most damaging at that time. The recession hit the railroads hard, forcing a quarter to shut down. What the railroads once reaped; they would have to repay. Congress had spent the previous 15 years handing out generous funds as rail stimulus which led to overexpansion. The 1873 recession amplified the railroads' ongoing financial issues.

Violence, death, and devastation started in Martinsburg, West Virginia, and then spread to other states. The damage totaled millions of dollars and included the destruction of locomotives, freight cars, and buildings. Riots left over forty men dead and hundreds wounded. Federal troops forced an end to the demonstrations forty-five days after its start.

Armory, Columbus, Ohio. ca. 1904. *Courtesy Library of Congress - Detroit Publishing Company*

From the disastrous results of the 1877 uprising, both the leaders of disgruntled workers and the public prepared for future events. Labor turned to strategic planning and invested in strike funds to strengthen walkouts. The public responded with a round of state militia armories centered in many Northeastern cities, and recognizable today by their familiar medieval castle facade.

### 1886-Haymarket Affair-Chicago

Labor organizers chose May Day to lobby for an eight-hour workday and bring attention to other labor grievances. Though not designed as a violent protest, it turned into one under strange circumstances.

At 10:30 AM, Chicago police broke up what had been a peaceful labor rally. As the protesters dispersed, someone threw a pipe bomb that exploded in front of a line of policemen. The explosion and subsequent dust cloud created panic, followed by a heavy volley of shots fired, mostly by the police. The shooting lasted less than five minutes. In the aftermath, eight police officers lay dead or dying, along with at least four civilians. Investigators never determined the identity of the bomber, but they suspected the person to be a union organizer or a communist sympathizer, although an anti-union instigator is a possibility.

The judicial system reacted with an uncharacteristic American trial. That next year, the state of Illinois prosecuted, and later executed by hanging, four anarchists for their direct or indirect roles in the Haymarket rally. Prosecutors based those accusations on the tenet that their actions led to the death of one of the eight Chicago police officers. A fifth defendant committed suicide the night before his scheduled hanging.

### 1886-Great Southwest Railroad Strike of 1886-Arkansas, Illinois, Kansas, Missouri, and Texas.

Two-hundred-thousand workers organized under the Knights of Labor walked off their jobs in a labor dispute. The firing of a coworker was the last straw compounded by the union's frustration over low wages, long hours, and lax safety practices. The owner of the affected railroads, the noted industrialist and infamous robber baron (in some circles) Jay Gould, hired replacement workers, and according to legend made the following statement, "I can hire one-half of the working class to kill the other half." The violence included the usual carnage and sabotage of public and private property. Texas and Missouri had to call up their state militias.

The strike ended without worker gains.

### 1892–Buffalo Switchmen's Strike–Buffalo, New York

Railroad workers staged a walkout as a rally to support existing New York State laws that mandated a 10-hour maximum workday and set minimum-wage rules. The railroads countered the walkout by hiring replacement workers. Protesters retaliated by destroying millions-of-dollars' worth of private and public property. The mayhem included strikers placing dynamite landmines on train tracks.

The state sent in 8,000 militia from New York City to end it. After two weeks, the strikers agreed to return to work; however, the company blacklisted some of the union organizers.

### 1892–Homestead Strike–Carnegie Steel, Pittsburgh, Pennsylvania

The Amalgamated Association of Iron and Steel Workers (AA) marked an organized labor milestone by negotiating collective bargaining agreements with steel mills and setting up rules for pay and hours. The steel workers even prepared for the worst with a sizable strike fund. Nevertheless, the union pushed mill management too far. The company responded by locking out the strikers from the steel mill compound.

Carnegie Steel Company operations manager Henry Frick (1849–1919) hired non-union workers and turned the mill into a fortress to house and protect them. The AA reacted by arming themselves—intending to shut down the steel mill's operations.

When Pinkerton sent in mill reinforcements on a river barge, labor union forces with rocks, guns, and a cannon ambushed them from higher ground. Nine union members and seven Pinkerton agents died in the battle.

A week later, the Governor of Pennsylvania called in the state militia to quell the violence. The strike soon collapsed as AA union workers crossed the picket line.

The Homestead strike proved to be a public relations catastrophe for the steelworkers' union, which had hoped for a sympathetic reaction from the public. Soon after the strike ended, an anarchist gained entry to Henry Frick's office, shooting him twice and then stabbing him a few times—Frick survived (the attacker was not a steelworker). Another PR blunder occurred when union workers pummeled two captured security men in front of journalists. The strike was so disastrous, organized labor would not return to the steel mills for 45 years.

### 1897-Lattimer Massacre-Pennsylvania.

Three-thousand Eastern European immigrant coal miners went on strike upset over a long list of grievances that included low pay, discrimination, assignment to the dirtiest jobs, and forced dependence on the company store. Their strategy was to march in large numbers from mine to mine to bolster support from neighboring coal miners.

Outside of Lattimer, a local *posse* (mine management) met the mob before they reached the mine. The armed deputies presumably overreacted against the unarmed striking mineworkers killing 19 and wounding many more. State authorities called in 2,500 National Guard troops to restore order.

### 1902-Anthracite Coal Strike-eastern Pennsylvania

Represented by the United Mine Workers of America, 150,000 anthracite coal miners demanded a say in the operations of their industry, an 8-hour workday, and a 20% hourly pay increase. But this was not a great time for a hardline stance. At the time, the coal industry suffered from a massive over-supply of inventories while the labor market had a plentiful supply of replacement workers. The two sides failed to produce a deal, so workers went on strike. The usual militant response included murder and vandalism.

President Teddy Roosevelt threatened to send in the Army to take over the coalfields, but instead, the President assigned Commissioner of Labor, Carroll D. Wright (1840–1909), to investigate the strike. His commission found average wages matched company profits, but suggested the owners split the difference on the workers' demands. The owners granted their employees a nine-hour workday, and for many a 10% raise, but the company refused to recognize the union.

### 1913-Ludlow Massacre-Ludlow, Colorado

The mineworker's union demanded union recognition, a 10% pay raise, an eight-hour workday, better accounting for actual work performed, and more flexibility on housing and medical benefits. Statistics highlight the nature of their dangerous work. Miners died on the job at a rate of over 50 men a year in the Colorado coal mines and the miners believed they were pursuing reasonable demands.

The coal mine companies believed they could replace their largely eastern and southern European workforce with more of the same.

In a show of defiance, demonstrators set up large tent encampments of up to 200 families, positioned to badger potential strikebreakers. Within a week of the walkout, escalating violence forced the governor of Colorado to mobilize the state's National Guard troops. The two opposing groups harassed each other for the next several months. On April 20, impatient troops shot up the largest tent city near Ludlow, Colorado, killing 20 people, mostly women and children. The Ludlow Massacre and other local battles became known as the Colorado Coal Field War.

The aftermath of the 1913–1914 labor wars left many without work and permanently damaged the Colorado coal mine industry.

## 1921–Battle of Blair Mountain–Logan County, West Virginia

The Battle of Blair Mountain pitted workers from West Virginia's union coal mining counties against nonunion Logan County. The conflict began as 15,000 members of the *redneck* army (named for their red bandanas) marched into Logan County to convert the nonunion mines to union mines. They confronted 2,000 private security guards and the resulting civil war approached the scale of military combat. Weapons included countless guns, a fortified freight train, and planes that dropped dynamite bombs. Federal troops ended the conflict but not until an estimated 50 to 100 union miners and another 30-odd private and public security forces were killed.

## 1922–Herrin Massacre–Williamson County, Illinois

As part of the Williamson County Coal Wars, members of the United Mine Workers of America (UMWA) shut down every coal mine in the county except one, the Southern Illinois Coal Company. The small firm operated with 50 Irish replacement workers, protected by a handful of mine guards. Southern Illinois continued minimal coal operations to pay bills. Apparently, that was in violation of an agreement between the mine owner and the union. A jilted union mob retaliated by destroying mine equipment and firing weapons at the nonunion workers. Eventually, the union lured the strikebreakers into a trap under a supposed "surrender and we will let you go" deal, only to massacre nineteen. The slaughter included instances of the strikers approaching helpless *scabs* as they lay dying from their gunshot wounds. They then slit their throats with knives and urinated on them. History books tend to omit this strike, since those events do not bode well for organized labor's populist character.

✧✧✧

Organized labor believed that economic growth and industrial consolidation would give workers substantial leverage. But it did not. Strikes not only failed, as the previous examples show, they often involved terrorist-level violence.

Big labor did not fare well for several reasons. For one, many of America's horrific strikes occurred during or near economic downturns, notably the panics and depressions of 1873-79, 1882-85, 1893-94, 1896-97, 1907-08, 1920-21, and later the Great Depression. Bad timing led to two problems for labor. Before the days of easy money, failing economies depressed the value of currency leading to severe bouts of deflation and ultimately pay cuts. While workers seem to understand and embrace inflationary cost-of-living wage increases, workers do not always recognize its logical inverse, wage cuts triggered by deflation. Pay cuts led to anger and anger to strikes. Another problem for organized labor during economic downturns was that businesses suffered from sagging sales, so companies often welcomed the strikes as an alternative to layoffs. The workers got the blame while the businessman cleaned out his slow-moving inventory.

An additional major obstacle standing in the way of higher wages came from an excess of idle labor, a nearly endless supply of potential strikebreakers. They reduced the tactical power of a strike and made peaceful negotiations impossible. The supply of scabs originated not just from foreign immigrants, but also from domestic sources, particularly former slaves and their descendants.

When America outlawed slavery in 1865 with the Thirteenth Amendment to the US Constitution, millions of former African American slaves entered a mostly white competitive workforce. Despite newfound freedom, they still faced severe and deplorable discrimination, and despite a willingness to work hard for low wages, they had trouble finding good jobs. Of course, low-cost competition from overseas workers did more than its share to dampen the power of the labor movement.

During the nineteenth and twentieth centuries, many countries nurtured enough advanced agriculture to feed a population boom, but they lacked the latest *Western* technology needed to support a large industrial job base. Meanwhile, developing technology had built the steamship. The new generation of ocean-going vessels carried more passengers at higher speeds. With technology effectively shrinking the oceans, immigrants from first to third world countries, particularly Irish, German, Italian, Eastern European, and Chinese, flooded American job markets.

Labor also had to address their political radicalism. Extreme left-wing politics became entrenched in labor organizations. Socialists, communists, and anarchists resented the increasing income discrepancies between employers and employees. They even begrudged private ownership of businesses, as factories, railroads, and mines should have been the *peoples' property*. Many of those angry political ideals came from European immigrants bringing with them their brand of radical unionism.

Leading up to the FDR administration's inaugural term, both government and labor took a few strides toward easing the issues of surplus jobs and political extremism.

The federal government enacted laws to reduce immigration in an effort to lessen the unskilled labor surplus problem. The Chinese Exclusion Act of 1882 banned all Chinese laborers from entering the United States. In 1917, the U.S. added English literacy requirements for all adult immigrants. Congress' Immigration Act of 1924 (Johnson-Reed Act) limited the annual number of immigrants from any nation to that of 2% of the 1890 Census. Figure 27.1 charts the results of immigration reforms. The percentage of America's foreign-born population hovered at about 1% between 1900 and 1950.

For their part, organized labor toned down its political radicalism, distancing themselves from the far-left militants and shifting to the mere left-leaning Democratic Party. That political shift became critical to labor's acceptance by the more conservative mainstream America.

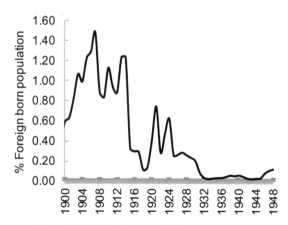

**Figure 27.1** Foreign born population as a percentage of US population, 1900-1948

Organized labor made other gains. Most notably, they became more inclusive. Worker organizations, traditionally made up of *white, Western Europeans,* now included African-Americans and other ethnic groups. Unions representing skilled (craft) and unskilled workers united to form a vast network of the *brotherhood* thereby creating a valuable culture of solidarity among workers. The ability to have the employees of a competitor's firm or another industry walk out in unity became a powerful labor tool.

Lost wages during extended strikes presented yet another fundamental problem for the labor movement. To persuade more workers to strike, union leadership needed to reduce the risk of lost income. The solution—build a strike fund. By charging union dues, labor could invest money into a fund to compensate at least a part of a laborer's lost income while they worked the picket line.

Nevertheless, to secure widespread collective bargaining agreements for higher wages, increased job security, etc., organized labor needed a government willing to grant those privileges not available in a free market. In 1933, workers' groups got their man when Franklin D. Roosevelt, labor's friend, became President. He had also surrounded himself with liberal bureaucrats enamored by pro-labor, progressive ideals. FDR's brain trust supported workplace issues, and they were sympathetic toward organized labor's struggles to strike effectively.

President Roosevelt delegated Frances Perkins as his Secretary of Labor. Perkins, the first woman appointed to a US Cabinet position, held the chief labor post from 1933 to 1945. Prior to working for the White House, the Secretary worked as a labor advocate in New York state government.

As promised, Roosevelt's pro-labor legislation started within his first one-hundred days in office. On June 16, 1933, the President signed into law the National Industrial Recovery Act (NIRA), an economic stimulus package designed to lift the economy out of the Great Depression. For the NIRA, FDR assigned Perkins as its chief labor architect. The pair influenced Congress to legislate statutes that guaranteed the right of workers to unionize, plus the bill allowed the President to set minimum wages and maximum workweek hours. As part of the labor package, the FDR administration created committees to promote *fair* labor conditions. After the passage of the NIRA, Roosevelt delivered a statement in which he said,

> History probably will record the National Industrial Recovery Act as the most important and far-reaching legislation ever enacted by the American Congress. It represents a supreme effort to stabilize for all time the many factors which make for the prosperity of the Nation, and the preservation of American standards.

Contrary to his optimism, FDR's policy did not pacify employer-organized labor relations. In one of the many worker disruptions following passage of the NIRA, Pacific coast longshoremen walked off the job in the early summer of 1934. They went on strike to protest the failure of their union to negotiate a *closed shop* policy with their employers. The harbor freight companies replied by hiring strikebreakers, which then led to a more aggressive, militant response by the strikers. The governor called in Guard troops, and along with security forces, they killed several protesters before what we now know as the West Coast Waterfront Strike ended.

The bigger blow to peace between business owners and a unionized American workforce came later that summer, eighteen months into FDR's first Presidential term. A massive strike highlighted the fact that despite having an ally in the White House, plus many in Congress, little had changed for the millions of struggling workers trying to gain control over their destiny.

On Labor Day, September 03, 1934, deep into the Great Depression, Southern members of the United Textile Workers (UTW) went on strike after cotton mill owners ignored their list of grievances. In response, the industry's sparse union membership sent out *flying squadrons* to coerce other mills to join the strike. The *mob-mentality* tactic worked, triggering spontaneous walkouts throughout the South. Momentum then spread to Mid-Atlantic and Northern states. Within a few days, nearly 400,000 mill workers from Georgia to Maine joined the protest, the largest labor strike in American history at that time. The protesters' list of demands included union recognition, higher pay, fewer working hours, and lighter workloads.

The Great Depression created dire economic conditions that negatively affected the labor movement. An oversupply of finished product, some pre-dating the stock market crash, meant that the mills were in no position to accept anything near to the protesting workers' ambitious terms. Meanwhile, communities and churches ignored the strikers' pleas for support. Out of work and on their own, the reality set in that the strikers would not achieve their goals. Frustration and desperation ended in hostile clashes between workers and mill management.

The rash of violent confrontations led local law enforcement agencies to police the strike. After only 22 days, the walkout ended, and the mills reopened. At least ten strikers, including six in Honea Path, South Carolina, and one police officer died. The failed walkout of neighbor versus neighbor left labor organizers blackballed from future mill work and broke the union in the heavy-textile Southern states.

Regardless of labor reforms backed by the power of the federal government, the Strike of '34 demonstrated the lack of progress during the previous six decades of tense conflict between labor activists and society. However, from those 60 years of walkouts, recent political events taught labor leaders a valuable lesson: organized labor needed stronger federal laws that could add bite to the lackluster power of a strike.

The Norris-LaGuardia Act of 1932 and the NIRA offered some relief; however, labor achieved a major victory on July 5, 1935, when President Roosevelt signed into law the National Labor Relations Act (known as the Wagner Act), which Congress specifically engineered to boost organized labor's power. The Wagner Act established the National Labor Relations Board (NLRB), created statutes prohibiting employers from interfering with union activities, and most significantly, it instituted a nationwide closed shop policy, which allowed a majority vote to force the minority to join a union and pay union dues.

The Wagner Act provided a major boon to organized labor. Until Wagner, labor unions lost virtually every battle. Strike after strike, entire industries refused to negotiate with labor. When unions appeared to have the upper hand, companies might close their doors or relocate. Instead of receiving better wages, an innumerable number of workers lost their jobs while hundreds died in violent strikes. Outside of work, unions earned a reputation as radical political activists.

After the passage of the National Labor Relations Act, labor unions' representation of American workers rose from 8% in 1935 to 27% by 1942. Labor unions, who had lost most of the battles, finally defeated big business and declared victory in the 60-year long labor war.

Democrats and union bosses gained huge benefits from the latest round of pro-labor legislation. In effect, Congress granted organized labor the power to control dues-paying workers. In return, Democrats gained a large, loyal voting bloc, plus a huge, perennial donation source via union dues. Labor's elite also fared well with rewards of lucrative pay, cushy jobs, and great power.

To its credit, the Wagner Act reduced excessive labor-related violence, and companies ceased punishing striking workers. The legislation did not solve the bulk of workers' problems as market power never vanished. Labor disputes continued at a furious pace negatively affecting wages through lost production and declining sales.

But politics was still the progressive's domain in the late 1930s, with more work to be done to increase membership and augment workers' leverage in a democratic-capitalist society.

## The Fair Labor Standards Act of 1938—The Good, the Bad, and the Paradoxical

Fifteen months into Roosevelt's second presidential term, Congress passed one final piece of New Deal labor legislation, the Fair Labor Standards Act of 1938 (FLSA)—another gift from FDR to organized labor. The FLSA strengthened anti-abusive child labor laws, introduced a federal minimum wage, and guaranteed premium overtime pay for hourly workers. While progressives still consider all three statutes an important step forward for working Americans, pragmatically the trio returned a mixed bag of results, more like the good, the bad, and the paradoxical.

**The Good** - Labor unions strongly backed child-labor reform for two reasons. First, they considered children to be low-cost competition to its dues-paying members. Second, and the stronger motive, organized labor's humanitarian opposition to child labor brought positive public-relation exposure.

Relative to other cultures, the US did not have a child-labor problem in the twentieth century; nevertheless, for those occasions when they arose, the new child-labor laws addressed a serious issue often in need of reexamination.

**The Bad** - The FLSA's real damage to a free society followed the enactment of the federally mandated minimum wage. Several states had approved minimum wage statutes just prior to World War I. At the national level, Congress passed a national minimum wage requirement in 1933, as part of the National Industrial Recovery Act. In the case of every one of those laws, state or federal, the United States Supreme Court declared a compulsory minimum wage unconstitutional, citing such a law as a violation of a worker's freedom. The FLSA's rules imposed a minimum wage of 25 cents per hour increasing to 39 cents in 1939, 40 cents in 1944, followed by a large jump to 75 cents by 1950. This time, as the nation's highest court politics had shifted to the left, the law remained intact.

Unions demanded minimum-wage regulations from their political allies to discourage the hiring of replacement workers, particularly minorities. For social reasons, political progressives believed that the statute gave workers a chance at a *living* wage.

Whether national or local, few American laws match the depravity of a minimum wage. As a government-backed price control, the law excludes the poorest of the poor from the job market. When the minimum hourly wage was 25 cents, workers with a market value of about 22 cents per hour probably earned a wage boost. However, the worker whose value was a mere 12 cents per hour would have to find work outside the federal law. Early rules governing the FLSA's minimum wage exempted smaller companies and some industries from the regulation. Nevertheless, the agency's grip on lower-wage earners expanded over time.

Another natural reaction by employers to the minimum wage laws was to cut an employee's weekly hours. Dora L. Costa (b. 1964), a research economist at UCLA discovered such a trend in her 1998 publication, "Hours of Work and the Fair Labor Standards Act: A Study of Retail and Wholesale Trade, 1938-1950." Professor Costa's data indicated a 5% drop in hours worked per week because of the FLSA minimum wage laws.

Minimum-wage laws remain popular among today's pundits, especially political liberals. One reason for its widespread support is that published studies rarely indicate any adverse problems related to mandated wage hikes. Politicians and analysts alike view them as something good for the economy with only minor negative side effects. Despite the passionate rhetoric directed at the need to boost the wages of the poor, the feds rarely make dramatic jumps in the national minimum wage. Historically, increases follow government-published annual inflation rates of 3%.

The trick to a *healthy* minimum wage is to reduce unfavorable wage and unemployment data to statistical noise. Thus, the ideal minimum wage requires precision, a *sweet spot*, an increase large enough to appear worthwhile, yet small enough not to set off any alarms.

The effects of a minimum wage, positive or negative, as is the case in many areas of Great Depression scholarship, eludes a true perspective. Common sense suggests that for some of the long-term unemployed, especially unskilled people living in poverty-stricken neighborhoods or Americans facing 1930s race discrimination, the minimum wage became an obstacle to good, steady work.

One other note pertaining to both then and now: the minimum wage dictates an hourly pay rate, not weekly earnings. Workers can augment their salary with additional hours or a second job. Either way, with experience, the majority can earn wages well above the minimum standard. Alternatively, forced out of the job market, people may turn to a life of government dependency. With the economy still in a slump, the minimum wage came at the most inconvenient time.

**The Paradoxical** - The overtime statute set the standard work week at forty hours—sort of. Instead of a flat maximum, as enacted by some European countries, the FLSA forced the employers of hourly paid workers to pay 1½ times base pay for all hours exceeding forty. Progressive politicians favored the concept of a shorter workweek to ease unemployment. In their view, three workers at forty hours per week made more sense than two workers at sixty hours per week.

The surprise reaction to the overtime statute covered by the FLSA came from union workers—a classic case of unintended consequences. For unions, the goal of shorter hours became a cornerstone of organized labor philosophy. Spanning over 50 years, some of the most famous worker uprisings featured the "forty-hour workweek" battle cry. Men died under the protest banner of a fair workweek. But the rule backfired. Market forces reigned over idealism as workers, instead of embracing the shorter workweek, clamored to get their *fair* share of overtime pay.

After the enactment of increased overtime pay, so many workers opted for fatter paychecks instead of fewer hours that unions crafted their own rules. Those new rules spread overtime evenly among union members so that a *privileged* worker did not benefit more than their union brothers and sisters. Here, supply and demand defeated the command economy, questioning the pragmatism of labor's idealism.

## Labor Conclusion

A rift exists between the wealth generated by free labor and the constraints sought by organized labor. For the bulk of private-sector workers, increased wealth does not come from big government-backed privilege, bureaucratic rules, and regulations, but instead, from a workplace that embraces increased productivity—an hour's worth of labor that delivers more goods or services. That path to higher production develops through a lean and flexible workforce. To that goal, profitable businesses depend on an employee base that adapts to innovative technologies, adjusts their hours to varying workloads, and moves labor from department to department, company to company, or sector to sector, depending on market conditions. Conversely, organized labor strives for status quo, and is often willing to restrain productivity through grievances to prevent change.

How did those differences correlate to the Great Depression? As with many other issues of the era, even a rough assessment is impossible to quantify. Yet, history and statistics offer a few insights. On the plus side, by curbing the frequency of violent strikes, Roosevelt's labor policies

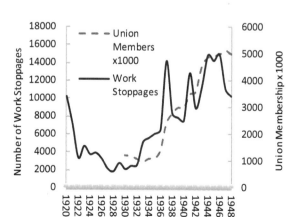

**Figure 27.2** Union membership 1930-1948 and the number of work stoppages, 1920-1948

must have had some positive effects on both the economy and society. On the negative side, following federal pro-labor legislation, the number of work stoppages increased (see Figure 27.2). The problem became so dire that if Congress had not intervened after the war then there might not have been an economic recovery. A later chapter discusses that crisis.

# 28
# Perpetual Jobs

FDR's administration and Congress established a wide variety of federal agencies designed to reduce unemployment. The two mega anchor programs were the Civilian Conservation Corps (CCC, 1933-1942), designed to help fund infrastructure programs, and the Works Progress Administration, (WPA, 1935-1943), created to hire workers for smaller infrastructure projects and to promote the arts. Several minor agencies funded local community-centered programs.

Proponents of the New Deal consider its jobs initiatives the cornerstone of high unemployment-busting legislation. Those pundits point to the number of accomplishments completed in such a short time, often from scratch. At a glance, the list appears impressive. Jobs programs built, repaired, or renovated 200,000 miles of roads, 3,000 bridges, 1,000 hospitals, and 20,000 schools. Thanks to the New Deal, art lovers enjoyed impressive murals. Hikers and mountain bikers still traverse thousands of miles of hiking trails built by the reemployed. Famous landmarks include LaGuardia Airport, the Triborough Bridge (renamed the Robert F. Kennedy Bridge in 2008), the Lincoln Tunnel in New York City; the San Francisco-Oakland Bay Bridge in California, Acadia National Park in Maine, and the massive projects funded by the Tennessee Valley Authority, which supplies flood control and electricity to five states.

Jobs programs produced staggering results. They also consumed staggering resources. In fact, New Deal public works programs employed three to four million workers each year during their peak period from 1936 through 1941—nearly 10% of the workforce.

The flood of anecdotal success stories does not correlate to the main accomplishment one would expect, as the Great Depression did not end until at least three years after the elimination of the last remaining program. In truth, jobs programs proved disastrous for the economy. The redistribution of labor did the recovery no favors and instead helped exacerbate, not relieve, the Great Depression.

That said, for being government sponsored, Roosevelt's team did an extraordinary job launching a wide variety of impressive projects. A lack of execution was not the problem. Conversely, it was no surprise that some agencies earned a lousy reputation. Noticing that there were too many idle workers on the government payroll, people played with their acronyms such as the WPA to stand for <u>W</u>e <u>P</u>utter <u>A</u>round. There were other problems such as targeting swing states in order to boost the election hopes of Democratic politicians.

Despite the earned criticism, those programs may have been the best run by the federal government, outside of national defense. But their success was relative. Jobs legislation failed because it shifted a large percentage of the economy's workforce from the private sector into the public sector. Even during the depression, employment needed to follow the market, providing what the people wanted, not what the elitist politicians and bureaucrats envisioned.

For example, in the early days of air travel, a depressed 1933 economy did not need a New York airport; Newark, New Jersey had one on the other side of the Hudson River. Infrastructure for automobile and truck traffic should have been local affairs, not federal. Destitute Americans could live without expensive art, music, and foot trails, or the well-written guides promoting national parks.

America's policy makers embraced many failed principals during the Great Depression—count job stimulus as another one and another lesson not learned. The prevailing thought among politicians and academics on economic downturns is that people are not spending and investing because they do not have the money, or their bleak fiscal outlook favors caution. Politicians respond to the downturn with their stimulus hypothesis, which suggests that unemployed people earning newfound income will spend that cash on purchases—breaking the recessionary log jam.

Part of that rationalization holds some truth. Economies will dip, and that dip is a lack of both producer and consumer confidence, causing people and businesses to throttle back spending. However, the economy always possesses a great desire to produce and spend, especially if suppliers can meet market demands. If left alone, austerity will run its course allowing economic growth to resume.

Self-defeating government *stimulus* programs yanked workers out of the private sector and then required increased taxes on the public to pay their salaries and overhead. Poor allocations of resources put on an impressive show, but they did not offer much relief and instead extended the misery of the Great Depression.

Scientists have a term for any magic mechanism capable of producing work without an energy source—a perpetual-motion machine. To counter the economic slump caused by the Great Depression, politicians' funding of massive *employment* initiatives created their own social version, a fantasy concept not unlike E.C. Escher's perpetual motion illustration *Waterfall*.

The buildup toward the war effort, beginning in 1940, signaled the end of Roosevelt's worker programs. World War II killed off the last straggler—the WPA, RIP–1943.

# 29

# Farm Aid

The plights of the hard-working American farmer symbolizes the tough times of the Great Depression. Nature's fury, especially the infamous Dust Bowl, farmers struggling with the market's quirks, and expensive economic reforms all made the tough job of farming more difficult. But does Great Depression farming live up to its dismal reputation, or does agriculture mirror manufacturing industries—despair mixed in with prolific production?

Charts 29.1, Wholesale Wheat Prices—1910-1932, and 29.2, Wholesale Cotton Prices—1910-1932, are two typical examples that highlight the dramatic price fluctuations of farm goods during the 23 years preceding the New Deal. The graphs reflect agriculture's volatility during that era which were triggered by a host influences such as war, new technologies, recessions, and government intervention.

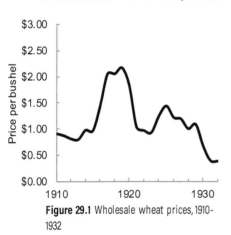

**Figure 29.1** Wholesale wheat prices, 1910-1932

The larger twentieth century agriculture problems originated from World War I. The First World War immersed a large part of the European continent into battle with many farmers trading their tools for rifles and trudging off to war. As a result, food supplies plummeted until American agriculture came to the rescue. Yet, in classic supply and demand fashion, the food shortage in Europe forced American food prices to rise. After the war, Europeans returned from the battlefield to their farms. American farmers now faced lower demand and with too much supply, prices plummeted, as did the number of farmers.

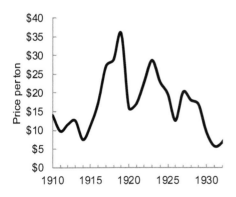

**Figure 29.2** Wholesale cotton prices, 1910-1932

The vibrant 1920s at least offered a temporary reprieve from falling prices.

In the meantime, innovation had been at work transforming farming as it had done with most industries, creating its own brand of *turmoil*. By 1929, farmers owned and operated one-million tractors, each with the ability to perform a multitude of tasks. In that same period, the number of horses declined by eight million, a sign of the productivity advantages of tractors over beasts of burden. The reduction in horses also freed up valuable pastures used to feed work horses. Farmers benefited from each new generation of trucks, along with better roads to drive them on, reducing the time to receive supplies and get their goods to market. The rural power grid expanded 350% from 1920 to 1929. Electricity powered irrigation pumps, auxiliary machinery, heaters, and provided a safer source of lighting for barns. Farms increased their use of commercial fertilizers by three times during the decade. Besides rapid transportation, early refrigeration helped cut waste. Industry, as they did for all occupations, supplied farmers with a variety of innovative, clever tools.

The Wheat Prices and Cotton Prices graphs show what is in part the effect of mechanization on farm goods' pricing from 1921 into the early 1930s. The next chart, which shows the decline in farm labor as a percentage of all workers averaged over a decade, also illustrates technology's ability to consolidate industries by leveraging productivity.

Figure 29.3, Changes in Farm Employment, charts the percentage of American workers engaged in farming from 1840 into the twenty-first century, providing an insightful illustration regarding the natural consequences of economic growth. In 1840, farms employed over 65% of American workers. That number fell consistently until the 1970s when the percentage dropped to near 3%. Note that the 1920s and 1930s were not unique, at least according to the graph. From the graph's perspective, even the Dust Bowl does not appear that historic. Despite declining labor, America ranked as the world's number-one producer of agricultural goods over that entire period.

Hunger was a national issue in the 1920s and a larger one during the 1930s. Politicians should have welcomed the trend toward lower prices. Plus, the surge in industrialization before the Great Depression increased the demand for workers to make, distribute, sell, and support new products. The streamlining of the agricultural workforce provided a

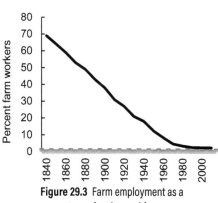

**Figure 29.3** Farm employment as a percentage of entire workforce

valuable source for that labor. Other farm-tech benefits—tractors, trucks, and electricity—made way for the hobby farmer, men and women who worked a full-time job while operating a small farm.

Despite a well-functioning market, the changing agricultural landscape made a major farm bill under progressive leadership inevitable. Farmers wanted help, and with a presence in every state, politicians wanted to help them. In fact, at the time of the Great Depression, farmers represented the largest percentage of workers of any industry. Farm lobbies and their progressive partners agreed that the free market was not a fair one.

Politicians' and farm advocates' primary goal was to help farmers avoid bankruptcy. Supporters looked at low-interest loans and technical aid but zeroed in on the low price of farm goods as the larger crisis, one that the federal government could correct with mandatory price supports. The fix defies common sense, yet conventional wisdom embraced it.

This is how price supports work—simply set a legal price for farm goods above the free-market price. For example, if wheat sells for $1.50 per bushel then write a law that sets the minimum price of wheat at $2.00 per bushel. The new price for a bushel of wheat becomes $2.00. In theory, if the wheat farmer made a $0.10 profit at $1.50/bushel with the generous price support the farmer should earn $0.60—big difference.

The fix might sound simple, but it plays havoc with supply and demand. For example, incentives generated by fixing higher prices will create an oversupply. Of course, the plan's architects had a solution—the federal government buys the surplus, which they either give away to the needy or, if they must, destroy. Others floated an alternative solution. If the federal government becomes overwhelmed with a surplus, which will happen, bureaucrats can pay farmers not to produce those items.

By the late 1920s, a growing class of intellectuals had bought into the lunacy of price supports. Bureaucrats at the US Department of Agriculture (USDA) also latched onto the concept with some added propaganda. They called the plan *price stabilization*. Farm supports were destined to be a massive welfare program, which would create waste, misdirect labor, and levy higher taxes. Unfortunately, this worst-case scenario was ready to come true when the American economy could least afford it.

President Hoover set out from day one to reign in unfettered industrial and agricultural capitalism. Four years of progressive farm legislation followed. The next administration, instead of placing the blame for the Great Depression on too much government interference, believed that Hoover did not enact enough anti-capitalist initiatives.

Roosevelt and his brain trust were not about to make the same mistakes. Yet, mistakes they made. In 1933, a commonsense ideology eluded the bulk of politicians and bureaucrats descending on Washington. Instead they clung to a far-left-leaning agenda. After all, capitalism had fatal flaws that needed mending. One of FDR's top priorities, which he assigned to his brain trust, was to implement a fast-track solution to what they perceived as America's continuing agricultural crisis. The President agreed with the establishment's viewpoint that the alarming rate of bankruptcy among farmers was both unacceptable and fixable.

First though, to understand America's farm crisis requires some much-needed perspective. In 1928, Soviet communists launched their first Five-Year plan, a socialist model designed to overhaul the economy. The next year, Soviet leader Joseph Stalin (1878-1953) mandated that the plan include collective farming. Economic disaster soon followed. Between 1931 and 1933, the political fallout caused by the authoritarian regime, combined with the disincentives of collective farming, led to the starving deaths of several million Soviets, primarily Ukrainians.

In contrast to the Soviet farm crisis, Americans produced massive quantities of agricultural goods. Many were hungry, but examples of Americans starving, which occur every so often regardless of economic conditions, did not arise because of the hard times. In fact, the United States produced so much food that throughout the Great Depression, America continued its reign as the world's largest exporter of agricultural goods.

The Soviet program led to a human tragedy, while the American effort would develop into an economic quagmire. The political elite wrapped both the Soviet's Five-year plans and FDR's New Deal in golden chains. Today, New Deal critics consider FDR's top farm bill a disaster. What did FDR do to draw the wrath of his modern critics?

President Roosevelt signed his pro-farm Agricultural Advancement Act into law on May 12, 1933. The fast-track program created the Agricultural Adjustment Administration (AAA), a replacement for President Hoover's Federal Farm Board. The agency fell under the control of the Department of Agriculture.

Roosevelt chose his future vice president Henry A. Wallace (1888–1965) to head the US Department of Agriculture (USDA). Secretary Wallace, as chief administrator of the USDA, would set the direction for the AAA. An agricultural academic, Wallace had studied farming his entire life.

Henry Wallace understood agricultural science at a high level, but his central-planning stance on farm economics turned the AAA into another New

Deal disaster. Hoover's farm bill, although ill conceived, at least had a meager budget that limited the damage—one-half billion dollars over four years. In contrast, FDR's version of Hoover's farm bill spent an average of $2 billion annually and another billion dollars per year for forestry programs.

The Agricultural Adjustment Administration's core mission was to reduce the supply of all agricultural goods, which would raise prices and afford farmers greater income, a ploy New Dealers believed would slow down the farm foreclosure rate and encourage farm families to spend. The scheme may have been what the political left wanted, but it was not what the American economy needed. First, American agriculture was not broken. As a response to modern industrialization, farmers had been increasing productivity and trimming their labor requirements for the previous century and for several decades following the economic panic. The 1930s farm economy fit that historical trend. Second, raising food prices by mandate during a time of heightened poverty goes against common sense. Finally, the AAA's economic policy took on a mighty foe—supply and demand. Undeterred by the enormity of the task, Henry Wallace mustered his army of administrators, along with those billions of budgeted dollars, and set out to make his contribution to economic stagnation.

Wallace devised a simple plan—buy up agricultural surplus. The problem was that in a highly regulated economy, which was the general direction of the New Deal, "too much surplus" is subjective and leaves plenty of room for interfering public servants to produce disastrous results. For example, in one of the farm bill's most infamous disasters, the federal government paid farmers to slaughter six million pigs, with over 90% of the carnage plowed into fields to rot as fertilizer. Another order had farmers plow under millions of acres of cotton fields. Those were two of the harshest examples, but there were many more from tobacco to peaches. Farm bureaucrats' overzealous attack on the free market was so reckless that the policy led to America's importation of enormous amounts of goods, including traditional American farm staples such as pork, beef, butter, wheat, and cotton.

The irrational practice did not last long. An outcry by critics from various fronts forced the AAA to halt their destructive tactics. Besides, there were other, *more acceptable*, methods to limit the supply of farm goods. For *Plan B*, farm bureaucrats turned to a process familiar to twenty-first century Americans—paying farmers not to farm. The new policy designed to drive up farm-goods prices was less barbaric, but it was just as foolish.

Despite progressives' objections to the agricultural free market, the system had been working. Farmers exited falling markets and moved to rising markets. The AAA pay-not-to-grow strategy changed those centuries-old tactics. The new policy, employed by many enterprising New Deal-era farmers, was to transfer their energies from farming to plotting—how to get the most government dollars from the least effort. As a result, the Agricultural Administration Act, as with so many of FDR's recovery initiatives, had more than its fair share of productivity-reducing unintended consequences. One adverse byproduct of the farm bill was the creation of the *accountant farmers*, who were harvesting federal dollars instead of crops or livestock. The practice led to idle acreage. In typical New Deal fashion, the hardest hit by the program were the nation's poorest—Southern sharecroppers.

On Jan 6, 1936, in a rare 9-0 judgment, the Supreme Court ordered the two-and-one-half-year-old Agricultural Adjustment Administration shut down. The judges did not fault the farm agency for its defective market practices, but found their directive to fund the program from taxes provided by food producers as unconstitutional. The SCOTUS decision to shutter the AAA proved to be only a minor setback for Wallace's farm agenda. Wallace re-named farm supports and moved them into another slot within the US Department of Agriculture.

American agriculture flourished after the introduction of the New Deal farm bills, but not to the credit of progressive farm legislation. Technology during the 1930s continued to transform the farm industry with advancements in electricity, tractors, trucks, fertilizers, refrigeration, and agricultural science. Those improvements dwarfed any efforts by Wallace's troupe of meddling civil servants. Yet, current conventional wisdom credits the AAA as a huge success story, and therefore, to this day, the idiocy continues as the USDA continues to pay farmers not to farm.

Wasted labor, wasted resources, and wasted goods—the New Deal farm legislation turned into another example of Roosevelt's brain trust putting the brakes on economic prosperity in the name of progress.

# 30

# The Predator Attacks its Prey

Three dynamic trends propelled the pre-Great Depression hyper-economy. The first was a boom in factory automation, which lowered prices and increased sales. Second, the utilitarian value of those products improved. Finally, industrialists launched a steady stream of new products for the American consumer, with many of those also following paths toward lower prices, higher volumes, and greater utility.

Most incredible, and often counterintuitive, is that profound technologies evolved without added labor—a positive trait during an era of full employment. To introduce new products in a craft-based economy requires the labor equivalent of *stealing from Peter to pay Paul*, which limits economic growth. Conversely, a conservation-of-labor economic model can reallocate resources to produce more goods.

Complex and widespread, the modern twentieth-century capitalist economy often masks the benevolence of industrialization, which although it included sacrifices, more than made up the difference with tremendous benefits. Burdened by their blind spot, progressives focused on the downside of expansion by developing their own warped perspective of the capitalist process. Even though they recognized some economic benefits of technology, progressives thought that unscrupulous business practices dominated corporate expansion. They believed that lower prices were not always technology or market driven, but a predatory ploy to eliminate competition. They believed that larger enterprises had an inherent cost advantage that they could use to drive smaller firms out of business. Once the cutthroat competition had eliminated the prey, the monopolist could raise prices and recoup the losses at the expense of consumers. For progressives, economic growth came down to greedy industrialists using unsavory methods for maximum profits.

Another concern pundits had toward unregulated capitalism was that, as a byproduct of the predatory model, vicious competition might suppress wages. In their view, lower wages reduced purchasing power which lowered factory output, causing a recessionary spiral. Conventional wisdom considers that cycle one of the major causes of the Great Depression.

Did big business place a detrimental burden on the 1920s economy that led to the Great Depression? It takes but a few examples to illustrate that the progressive's economic policy is a misguided interpretation of capitalism without a shred of deep economic thought.

Consider the economic influence of the iconic American automobile. In 1910, over 300 manufacturers produced automobiles. By 1929, near the economic peak and before the stock market crash, only about 75 automakers remained. All totaled, several hundred automakers folded, including the many startups that lasted but a few years or less. Previous chapters list several other industries in which consolidation produced similar dramatic results. To the anti-capitalist, the American system of manufacturing caused the demise of thousands of companies and the dismissal of millions of workers. Plus, tycoons such as Henry Ford became extremely rich.

Were the economic rewards worth the chaos? The answer is yes. Consider the Ford automobile. The price of the entry-level Model T runabout plummeted from $900 in 1910 to $265 in 1923. Ford's midlevel Tudor Touring sedan price dropped from $585 in 1924 to $500 in 1928 despite dramatic model upgrades accompanied by modest inflation. Price drops were impressive but here is the critical statistic: annual American automobile sales increased from 300,000 in 1911 to 1.2 million in 1921 and to over 4 million in 1929. In less than a single generation, US auto sales grew by a staggering 1400%. And it gets better. Producers did not just feed buyers bargain basement versions at reduced prices. A 1928 Ford Tudor compared to the 1923 Ford Tudor was larger, more powerful, easier to drive, more comfortable, superior handling, and safer—more car for less money.

Rich consumers did not account for the prodigious growth in automobiles. Instead, the beneficiaries were buyers spanning a broad range of income brackets. Nor were automobiles an anomaly, at least according to the US Census Bureau. The number of trucks produced swelled from 148,000 to 770,000 between 1921 and 1929, a five-fold increase in annual production.

In 1910, only about 3,000 farm tractors worked American farms. By 1929, farmers operated 827,000 tractors.

Aircraft sales rose from 437 in 1921 to 6,200 in 1929. Much of the growth came from the fledgling airline passenger industry.

Appliance makers sold 3,000 refrigerators in 1921 compared to 890,000 in 1929. Refrigerators represented one of the many consumer goods that were once only available to the rich.

The number of AM radios, introduced in 1921, totaled well into the millions by the end of the decade.

The textile industry introduced the first synthetic fiber, Rayon. The production of steel and the generation of electrical power, critical components to a host of industries, both dramatically increased during the 1920s. The list goes on and on as the breadth of innovative technologies covered a profound number of businesses.

Those results did not impress the progressives who, with newfound power, placed themselves in charge of American industry and created one of the most despotic initiatives in the nation's history—the National Recovery Administration (NRA).

The NRA, an industrial version of the Agricultural Adjustment Act, fell under the National Industrial Recovery Act, launched in June 1933. Congress formed the NRA arm to rescue an ailing economy by reducing unemployment, increasing wages, and improving industrial efficiency, or as Title I of the NIRA, with the NRA in mind, stated,

> A national emergency productive of widespread unemployment and disorganization of industry, which burdens interstate and foreign commerce, affects the public **welfare**, and undermines the standards of living of the American people.

Promoted by FDR's brain trust as a signature New Deal program, the agency would have unprecedented authority to police the free market through a partnership of industry, labor, and the federal government.

Some in Congress expressed their concerns. In response, the bill's final version came with a two-year trial-period provision. Still, the NRA became Roosevelt's most disastrous attempt to revive the economy, accompanied today with a dismal reputation. Even the bulk of New Deal enthusiasts condemn the initiative.

FDR assigned Hugh S. Johnson (1881–1942), a former army officer, lawyer, and businessman to lead the group. General Johnson, as he was known, helped draft the legislation from experience he had with President Wilson's War Industrial Board (WIB) (July 8, 1917–Jan 1, 1919). Johnson had served as the liaison between the US Army and the WIB.

The cornerstone of the NRA was its *Codes of Fair Competition*. To develop the codes, the program's architects organized business sectors into trade associations. Participation in the economy-boosting code committees was voluntary. To protect NRA members, Roosevelt had the DOJ suspend prosecutions against monopolistic practices, since the NRA guidelines violated antitrust laws by encouraging cartels.

NRA administrators designed standard templates and then directed the committees from each trade association to develop their own set of customized codes based on those templates. Once reviewed and signed by FDR, the Codes of Fair Competition became law. General standards included:

- a ban on child labor—under 16 years old
- an equal-pay-for-equal-work policy for women
- a directive that employers recognize labor unions and their right to collective bargaining
- a warning that larger firms should not drive smaller firms out of business
- a warning not to monopolize
- the establishment of rules governing industrial committees

Customization of the code included:

- minimum wages per industry—often tailored per geographical location
- maximum weekly working hours
- maximum production quotas
- standard business practices
- process standards
- product design standards

For a final touch, the NRA adopted a dramatic logo, a blue eagle, with the slogan, "We do our part." To the New Dealers, the blue eagle symbolized a patriotic commitment toward ending the business slump. The blue eagle stamp, displayed on American-made products, would show a patriotic compliance to NRA regulations by that business.

The NRA started with a burst of optimism. Several cities held large parades. Organized labor embraced the bill believing they were witnessing a new era for the working man. General Johnson tirelessly campaigned on behalf of the NRA. For his efforts, Johnson's hard work made him Roosevelt's most celebrated New Dealer. The NRA's lead man even made the cover of *Time Magazine*. Industries eagerly signed up beginning with the *big ten*—automotive, shipbuilding, textiles, etc.—with all onboard by mid-September 1933. The number of recruited trade associations totaled 557. Proud members displayed the NRA's blue eagle on their products and at their businesses.

Enticed by a price floor that might shield them from low-cost competition, many businessmen embraced the NRA. Henry Ford, an earlier proponent of the bill, changed his mind and proved the most notable exception. The Ford Motor Company already paid high wages. Plus, Mr. Ford did not agree with much of the NRA's language and refused to sign up. His rebellious act caused the loss of government contracts and the scorn of General Johnson, but because price-fixing codes did not hinder the automaker, Ford could skirt the rules and undercut the competition's prices and so did well.

Henry Ford's position was ahead of the curve—the NRA euphoria did not last. Complaints came from lots of sectors. Consumers protested the higher prices they had to pay for goods. Businesses were selective when choosing which codes they would follow or not follow, often adhering to price-fixing rules while ignoring labor policy. Labor unions did not gain the level of power they expected. Workers from all industries complained that the workweek limits amounted to a pay cut.

NRA codes hit black laborers harder. Minimum wage rules removed many from the workforce and prevented others from entering it. In addition, labor unions that were racist could use their influence with the NRA to exclude blacks from their ranks. Union racism was no surprise. As the economist and historian Walter E. Williams (b. 1936) observed, "Labor unions have a long history of discrimination against blacks." Williams noted two protests against labor unions by famous black activists. Frederick Douglass (1818–1895) in 1874 wrote an essay titled "The Folly, Tyranny, and Wickedness of Labor Unions" and decades later in 1913 Booker T. Washington (1856–1915) published an article, "The Negro` and the Labor Unions."

The once-enthusiastic bloc of business leaders even turned against the NRA. The mountains of new legislation made it difficult for businesses to determine which practices were lawful and which were unlawful, and for good reasons. Within months, the NRA produced more rules and regulations than the entire federal government had created over the previous 144 years.

Even Roosevelt's golden boy, General Johnson, ruined his once stellar reputation. An extramarital affair, public drunkenness, an abrasive personality, and the poor handling of a California labor strike proved the NRA's boss a PR nightmare for Roosevelt. FDR fired Johnson in September 1934. The President though avoided negative fallout as Congress received the bulk of public outrage.

The real problems with the NRA were more basic. Industry had been growing and changing at a prodigious rate. Even the 1930s were bursting with an incredible number of new and evolving technologies. The NRA's widespread use of mercantile bureaucratic standards stood juxtaposed against the tried-and-tested growth model of a free market. The attempt to freeze production methods not only helps explain the continued economic slowdown but also speaks volumes about the naive economic insight of FDR's brain trust.

The NRA regulations hit small businesses the hardest. Their survival depended on flexibility, innovation, and sometimes lower costs—all strategies inconsistent with the NRA's codes. General Johnson had no sympathy for them as he blamed the *little guy* for paying the lowest wages—in his view the cause of the Great Depression. The uproar over treatment of the small businessman led FDR to create the National Recovery Review Board, headed by the famed lawyer and left-wing ideologue Clarence Darrow (1857–1938). Darrow submitted three reports; all were critical of the NRA's handling of small business grievances. In response to his input, Roosevelt fired Darrow.

Despite its constructive reputation, the bulk of the New Deal produced harmful economic results which prevented a timely recovery and the National Recovery Act was especially brutal as it brought the nation to an oppressive place seldom seen in American history. The decades of the 1920s and 1930s marked a profound global shift in economic thought. The Communist Russians and the Fascist Italians took the lead, building regimes based on anti-capitalist platforms. Left-wing social reformers worldwide saw those types of command economies, run by benevolent leaders, as antidotes for capitalism's fundamental flaws. Or so it seemed.

Of course, authoritative state goals later clashed with those trying to exercise freedom of choice, forcing the Soviet and Italian totalitarian rulers to commit atrocities against their own populations for the good of the state. Despite some parallels in shifting economic policies, the United States was not the Soviet Union or Italy, and Roosevelt was not Stalin or Mussolini. Regardless the New Deal's light version of nationalizing industries (the NRA) sent the federal government in a chilling direction.

The regulations set by the NRA committees had a suggestive tone to them; however, once the President authorized the fair-competition codes as law, the executive branch had the newfound authority to enforce the codes as laws. A firm's officers became potential lawbreakers if their business did not meet a regulation—for example, charged too low a price or failed to perform a required operation as outlined by a private committee. Particularly disturbing was the breadth of the blue eagle's dominion. The library-sized fair-competition codes covered 23 million workers (40% of the labor force) and several thousand businesses.

The NRA administrators took enforcement of the immense volume of codes seriously. The executive branch hired thousands of NRA agents to police laws that business leaders created to hamper the effectiveness of their competitors. Predictably, the NRA became a hub of government abuse.

NRA violators did not have to be morally unlawful to feel the agency's wrath. For one, businesses had to understand a huge number of codes. But the larger issue often came down to a decision to either follow the code and avoid the NRA's wrath, or skirt the codes and remain competitive. In FDR's new-world vision of government-sponsored cartels, it would be small businesses whose existence depended on flexibility that paid the higher price for the flawed initiative.

Under authority of the NRA, the army of federal agents indicted hundreds of small business owners. The few who protested that the accusations against them were baseless were charged with code violations and sent to prison. The typical allegation: charging customers low prices.

The mistreatment of American business owners extended past the indictments. At a time when businesses struggled to survive, harassment by the NRA police could ruin or even close an enterprise by disrupting business and forcing legal fees. Fortunately, the devastation did not last long.

Regardless of the NRA's formidable foothold into the economy's private sector, the initiative ran into serious trouble. The complexity of the program caused former business allies to withdraw support, while the ineptness of NRA management turned politicians against it. Enough was enough. On May 27, 1935, less than two years after its inception, the Supreme Court issued orders to shut down the NRA. In a dramatic 9-0 decision, SCOTUS declared the NRA's method of code authorizing unconstitutional. The court ruled that it was Congress's responsibility to write laws and not that of a federal agency. SCOTUS also challenged the NRA's claim that the agency operated under authority granted by the Constitution's commerce clause. The justices' unanimous judgment sent a strong message to the President that he had pushed his constitutional authority too far.

Four days later, an infuriated FDR took to the airwaves and lambasted the Supreme Court for shutting down his pet New Deal initiative. In a not-so-congenial fireside rant, Roosevelt scolded the high court for applying, "the horse and buggy definition of interstate commerce." The President was not finished. FDR followed up his assault on the justices with his dastardly plan to pack the Supreme Court with like-minded judges to avert similar decisions.

The resolution to shutter the NRA ended one form of the government's intrusive meddling, but the political battles were not over. Intimidated by an enraged FDR, during the next round of New Deal legislation, the Supreme Court allowed more robust labor laws to pass when they resurfaced in the National Labor Relations Act and Fair Standards Labor Act. As for the business world, FDR had a backup agency he could use to punish industrial excellence—the Department of Justice.

# 31
# Punished for Excellence

By 1938, as the public, Congress, and the Supreme Court grew weary of the New Deal, the economy might have had a *window of opportunity* for a long overdue recovery. Unfortunately, FDR's brain trust brought in their *closer* to assure the three-plus term president would never witness a single moment of good economic times during his administrations. The progressive's champion was the US Department of Justice's revamped Antitrust Division led by Thurman Arnold (1891–1969). Arnold, yet another legal-professor-turned-member of Roosevelt's brain trust held the post from March 1938 to 1943. His organization more than quadrupled its mission to harass American businesses by enforcing the Sherman Act.

Antitrust legislation holds universal appeal. Many believe it is government's job to prevent ruthless businesses from forming monopolies. Thurman Arnold agreed. Arnold considered monopolies to be bottlenecks to American prosperity. In reality, the Sherman Act and subsequent antitrust bills are rotten to the core. Their punishments are harsh, details of the laws are vague, and they have tremendous power to dissuade the pursuit of excellence. Section 2 of the Sherman Act states:

> Every person who shall monopolize, or attempt to monopolize, or combine or conspire with any other person or persons, to monopolize any part of the trade or commerce among the several states, or with foreign nations, shall be deemed guilty of a felony…

But what is a monopoly? There is no simple answer. In fact, complete command of an industry is a rare feat. Then what percentage qualifies—90%, 51%, 40% in a crowded field? To this day, the law offers no clue, but its legal ambiguity persists. For example, how does Sherman define competitors? Was the expensive Duesenberg a competitor of General Motors' lower-budget vehicles? Is Budweiser a beverage, an alcoholic beverage, a beer, a lager beer, or more specific, an American lager? How about packaging—does it matter if the *King of Beers* sells their product in kegs, bottles, or cans? Do similar imports count as competition? Does Sherman apply to regional monopolies?

The time factor reveals another flaw since antitrust law lacks clear statutes of limitations. Prosecutors have the authority to go back several years or more to find corporate misbehavior. This ambiguity exercise goes on and on.

The gravity of Section 2 would seem to owe company officials some clarity. Instead of fixing a poorly designed law, legal architects turned to jurisprudence as the antitrust judge. In the court's view, there are two types of monopolies—predatory (bad) and natural (good). Predatory monopolies increase production and lower prices, which eliminates their competitors or forces them to submit to a buyout; whereas, natural monopolies increase production and lower prices, which eliminates their competitors or forces them to submit to a buyout. Yes, both the bad guys and the good guys rely on identical mechanisms to monopolize.

Nevertheless, the legal community maintains that they know how to differentiate between the two—but how? CEOs do not distribute edicts encouraging employees to use whatever despicable means to crush competition, with an all-expenses-paid vacation to Hawaii as a reward for the most aggressive. Instead, every large business on their way to dominating their market relies on thousands of innovations, thousands of internal directives, and millions of transactions. To establish a solid case, all federal prosecutors need to do is to cherry-pick the juiciest *abuses* of the millions of transactions that occurred over the years and then partner with a left-leaning judge to concur. Consequently, a legal team with enough manpower can concoct a legitimate antitrust case against any dominant company. As a bonus, prosecutors can use fuzzy statute of limitation rules to take their time—after all it's only the taxpayers' money.

How detrimental is industrialization? Remember the early days of the automobile? Hundreds of manufacturers produced thousands of automobiles. By the late 1920s and 1930s, dozens of manufactures produced millions of automobiles. It is obvious who benefited from the turnabout—lower volumes were for the rich, higher volumes for the masses. There were other differences between craft and high-volume automakers. Early autos suffered from poor quality, which meant higher maintenance costs. They were also under-engineered making them difficult to drive, thus many car owners needed a chauffeur. Hundreds of other industries have similar stories. Whether those industries had ten, five, or even one dominant player, consumers benefited.

Whether it was back in the 1930s or the twenty-first century, somewhere out in progressive jurisprudence land there are monopoly lines drawn that business leaders do not want to cross. They vary from prosecutor to prosecutor, judge to judge, and day to day. Businessmen should be aware; just getting close to that line is risky. Progressives did their propaganda homework. Charges such as monopolistic behavior, predatory practices, restraint of trade, exclusionary contracts, and price fixing infuriate American citizens. Of course, in typical antitrust fashion, the law does not provide legal explanations for any of those *crimes*. Thus, Section 1 of the Sherman Act, stated below, throws more caution toward those attempting to achieve excellence.

> Every contract, combination in the form of trust or otherwise, or conspiracy, in restraint of trade or commerce among the several states, or with foreign nations, is declared to be illegal.

In some circles, antitrust violations approach capital offenses. Thurman Arnold took advantage of public sentiment against monopolists by arresting offending company officers, fingerprinting them, and publicizing their cases to humiliate them. Nobody saw jail time though. Following tradition, prosecutors can make it all go away by exchanging criminal charges against company officials for large sums of money via fines or a civil lawsuit.

All that private-sector money making its way to federal coffers made Roosevelt a trustbuster believer. During his first term, FDR did not aggressively pursue trust busting. In fact, through the NRA and the AAA, his administration encouraged cartels by creating laws that promoted price fixing. Only after 1935, with the NRA and AAA ordered shut down by the Supreme Court, did the Antitrust Division expand their antimonopoly operations. To this day, mainstream pundits appear unfazed by the hypocrisy.

There are no means to determine the amount of economic damage caused by aggressive trust busting, but the fear of appearing to monopolize is out there. To the courts, even approved patents by the US Patent and Trademark Office do not afford companies the right to benefit from their own innovations. The vagueness of antitrust laws, which does not indicate where the do-not-cross monopoly line is, places corporate officers in a problematic position. The best way to avoid scrutiny by the DOJ is to back off on American exceptionalism and slow down innovation. Charge higher prices. Produce lower volumes. Antitrust legislation is one more reason sagging sales and high unemployment rates masked the benefits of an extraordinary industrial era as manufacturers continued to modernize throughout the 1930s.

Federal prewar antitrust harassments may have gone unnoticed, except for the most conservative critics, if not for one major blunder. In 1938, Arnold sent his goons after the Aluminum Company of America (Alcoa), a case that highlighted the stupidity of antitrust.

## Alcoa

Aluminum is the most common metal in the earth's crust—over 8% by mass; yet, due to its affinity for oxygen, the light metal does not exist in an elemental state. In fact, man could not extract quality aluminum until 1827, and that milestone required an expensive chemical process. For the next 50 years, because of high costs and low production, aluminum's value approached those of precious metals.

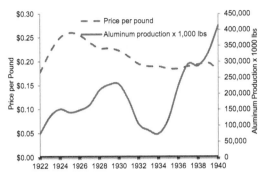

Figure 31.1 - Aluminum production and prices, 1922-1940

A major commercial breakthrough toward affordable aluminum arrived in 1886. That year, two young chemists, an American, Charles M. Hall (1863–1914), and a Frenchman, Paul Héroult (1863–1914)—yes, they share those same dates— independently extracted near-pure aluminum from alumina, a compound derived from bauxite. Both men used similar electrolysis processes. Avoiding a battle over discovery rights, the process became known as the Hall-Héroult electrolytic process. Incredibly, to this day it is the only method used to produce commercial virgin aluminum.

Charles Hall knew his breakthrough was something big, as did his wealthy investors. In 1888, Hall and his partners founded the Pittsburgh Reduction Company, which later became the Aluminum Company of America, and finally Alcoa. Hall obtained a US patent for his namesake process in April 1889, while Héroult gave up on aluminum and shifted his attention to steel.

Despite aluminum's potential and the advantage of ironclad patent protection, the young company encountered formidable obstacles. For one, the Hall-Héroult process required prodigious amounts of electricity. With the late nineteenth-century electric power industry in its infancy, electricity was dear and expensive and therefore so was aluminum. The new industry also lacked infrastructure. The extensive scope of the startup's business model included mining, smelting, material sciences, finishing, some fabrication, and distribution. Bauxite, the only practical source of aluminum ore, occurs only in tropical areas, forcing the Pittsburgh Reduction Company to build a global organization.

Another problem—in its natural state aluminum is soft and weak; thus it lacks industrial value. That dilemma forced Alcoa to invest into the emerging science of metallurgy in order to develop unique alloys and special heat-treat methods that increased aluminum's durability. Finally, starting out, the Pittsburgh Reduction Company had a minimal market. That was not the case for other corporate giant startups.

When Carnegie launched his empire, based on the low-cost Bessemer process, steel had been around for 4,000 years with potential customers awaiting affordable versions. Standard Oil's kerosene supplied an alternative to a vanishing supply of whale oil. Thomas Edison unleashed two novel technologies—the incandescent light bulb and the electrical power grid, both futile on their own, but combined, they changed the world. Edison and Westinghouse soon added the electric motor, further enhancing their value. Charles Hall had a product with enormous promise but with no real market waiting in the wings. In fact, when Hall built his business, inventors had not yet developed two major future applications, automobiles and airplanes.

Alcoa endured, despite the obstacles. They established a significant industry from scratch and then operated as its near-sole player for 40 years. Charles Hall's landmark discoveries turned into a true economic game changer.

Aluminum made the United States economy stronger and Americans richer. The electric power industry used thousands of miles of aluminum wire for long-distance transmission cables. Automobiles, trucks, buses, and trains used large amounts of aluminum to increase efficiency and reduce corrosion. The modern airline industry was totally dependent on aluminum—to name just a few.

The aluminum producer functioned as a monopoly, but not by design. Instead, high startup and operating costs, along with a still-developing market, discouraged challengers. Nevertheless, they dominated the market. Consequently, their success attracted the fury of the malcontents at the DOJ who were out to destroy America's manufacturing greatness.

The federal government first harassed Alcoa in 1914. That first generation of trustbusters wanted to temper monopolists' power by reminding them of who had the ultimate authority. The aluminum producer survived those attacks thanks to the Supreme Court's 1911 *rule of reason* doctrine, which stated that only abusive monopolies should be prosecuted under the Sherman Act. Unfortunately for the future of American prosperity, the next group of trustbusters, FDR's DOJ thugs, sought to destroy all monopolies regardless of what the rule of reason stood for. Alcoa would be its number-one target, with a goal to dissolve the aluminum giant.

In June 1938, Thurman Arnold's Antitrust Division stepped up its ongoing litigation into Alcoa, which ended with a historic, two-year trial comprising 58,000 pages of testimony. The court's decision came in March 1942, which found Alcoa innocent of antitrust violations. The government appealed, sending the case to the Supreme Court, which in June 1944, kicked the case down to the United States Court of Appeals to be judged by the Second Circuit. The Second Circuit overturned the lower court's ruling in March 1945, siding with the DOJ that Alcoa was too efficient for meaningful competition and therefore violated the Sherman Act. Instead of dissolving Alcoa, the intent of the DOJ, the court ordered that the government keep an eye on the corporation for five years—a mild reprimand.

In a way the system worked, as some justice prevailed, but the landmark decision exposed the many follies of antitrust. More than faults, the Alcoa case demonstrated that the Sherman Act serves no positive purpose in a free-capitalist economy.

Figure 31.1 charts aluminum production, dominated by Alcoa from 1922 to 1940. Notice, that by 1940, US production of aluminum had doubled from its 1929 former peak. Meanwhile, from 1925 on, prices declined. What more could a struggling economy ask for?

In a challenge to the DOJ, Alcoa claimed that they had competition. Of course, there is always competition for consumer dollars, but several foreign companies also produced virgin aluminum. More relevant was the vigorous scrap aluminum industry competing against Alcoa which the DOJ and the court rejected as *true* competition. People today appreciate what government prosecutors did not understand—the value of recycled aluminum. Domestic competition arrived in 1937 from Reynolds Aluminum, and 10 years later, Kaiser Aluminum entered the market.

Another noteworthy element of the case is the 58,000 pages of testimony, which despite the huge resources consumed by Roosevelt's henchmen, revealed no smoking gun to convict the defendant. Yet, all that litigation cost Alcoa millions of dollars.

The most damage came from the DOJ's challenge to the rule of reason doctrine. The decision sent a warning to any company desiring to increase production and lower prices. As Judge Learned Hand (1872–1961), representing the Second Circuit, wrote in his opinion: "Nothing compelled it to keep doubling and redoubling its capacity before others entered the field…" Later, future Federal Reserve chairman Alan Greenspan (b. 1926) weighed in on the *United States v. Alcoa* fiasco noting, "Alcoa is being condemned for being too successful, too efficient, and too good a competitor."

Roosevelt's antitrust policy might have lacked the destructive *tour de force* of its New Deal compatriots, but the brain trust's twisted socialist ideology toward monopolies captures the essence of progressives' ludicrous economic views.

## Conclusion

The New Deal and other FDR initiatives placed an enormous drag on the economy. Yet, the prodigious power of American industrialization persevered. In a bout of irony, while mainstream academia focused on what they believed were capitalism's failures, an economy emerged, far more productive than the former high mark created by manufacturing industries during the Roaring Twenties.[1]

---

1

# Part Five
# America Turns Modern

# 32

# The Nature of Invention

**The Great Inventors**
Today's train-wreck scholarship that dominates the Great Depression stems from academia's often shortsighted view of American industrialization. Looking back to the eighteenth and nineteenth centuries, historians believed they had a reasonable grasp of technology and the factory system. In those earlier days industrial pioneers delivered a tighter synopsis of technological progress: an inventor, an invention, a brief description of how their invention worked, followed by its dramatic social impact. Consider some examples.

- Steam Engine–James Watt–factories, locomotives, and steamships–1776
- Cotton Gin–Eli Whitney–a boost to cotton processing–1794
- Reaper–Cyrus McCormick–farming–1845
- Safety Elevator–Elisha Otis–the practical skyscraper–1853
- Telephone–Alexander Bell–instant long-distance communication–1876
- Light Bulb–Thomas Edison–light–1878

Each of those episodes, and many more, presented a rich story, often revering a humble individual with big dreams and an impossible challenge. The protagonist then triumphed through a combination of sweat, genius, and luck, capped with a few enduring anecdotes that chronicled their struggles. Some earned fortunes and kudos from their peers. Others suffered the scorn of their critics, ridiculed for their failures. The most unfortunate died steeped in debt as the next generation of tinkerers and businessmen reaped less-deserved financial rewards. Engineering-hero tales make history more concise and more interesting. The Currier and Ives-like industrial era lasted maybe a century following the Industrial Revolution. After that, things got complicated.

During the Gilded Age, manufactured products increased in complexity. Subtle innovations replaced game-changing inventions. Dozens or even hundreds of specialists organized in teams superseded the *genius* visionary and his or her capable apprentice. The inventors did not disappear, instead they became less conspicuous. Huge, faceless corporations with well-funded research labs took the place of the get-your-hands-dirty mad scientist. To the academic, a sea of advancements toward modernization overwhelmed the senses—from a few, to dozens, to hundreds, to thousands within a short period. Most disheartening for conventional wisdom was that dastardly big business became the heart and soul of progress.

As for documenting progress, consider that most modern inventions consist of a long, complex chain of innovations. Along the development process a few inventors might achieve fame, but the vast majority, the 99-plus%, toil in ambiguity. To further complicate histography, a fuzzy relationship exists between a product's publicity and its economic impact. Students of history, even with an economic or technical background, planning to solve the problem with data will find it impossible to quantify increases or decreases in economic wellbeing related to anything technological. Then consider its enormous scope. After the Industrial Revolution's big bang, technology followed its own infinity-bound expansion.

To gain perspective into capitalism's role during the Great Depression requires narratives and anecdotes instead of general *wellbeing* statistics. A speculative process might be too unorthodox for the purist, but the large volume of evidence is *in the details*, not in traditional monetary-based statistical methods.

Thomas Edison and the incandescent light bulb serves as an excellent primer to demonstrate the extensive gap between scientific discovery and an invention's meaningful contribution to the economy. Edison and his light bulb are well known, while the design and development of the pre-Great Depression phenomenon is surprisingly complex.

## Thomas Edison

In grade school we learned that Thomas Edison invented the light bulb. In fact, Edison's claim to fame is that he made significant contributions in the development of the incandescent light bulb.

Long before Edison's time, in 1809, an English scientist, Humphry Davy (1778–1829), wired a charcoal strip (carbon filament) to a battery (at the time, a cutting-edge technology) to create light. When he flipped the switch an electrical current flowed through the high-resistance, charged carbon wire creating enough heat that it produced a dim glow. Davy's famous experiment lasted a few moments, yet the brief experiment inspired a few inventors who recognized an opportunity to combine the valuable luxury of light with the revolutionary science of electricity.

After seven decades, more than a dozen inventors, and some twenty patents, two men closed in on a functional electric light bulb—an Englishman, Joseph Swan (1828–1914), and the American, Thomas Edison. In 1878, Swan demonstrated a lamp glow lasting 13.5 hours. The next year, Edison achieved a bulb life of forty hours. That clinched it. Edison received the credit for the *historical invention* and is the most recognizable name associated with the development of the electric lamp. But to get to that point, the Wizard of Menlo Park had to experiment with thousands of filaments in his quest to find an ideal material. By 1880, his *99 percent perspiration* efforts paid off—a carbon/bamboo composite filament that burned for an impressive 1,200 hours (50 days).

Swan and Edison made monumental breakthroughs, but they were still a long way from the practical light bulb. The ideal lamp filament, the key to light, must find a delicate balance—it needs to cook to extreme temperatures, yet without burning up. Early inventors experimented with two options, carbon and platinum. Carbon gained favor for its lower material costs. Nineteenth century inventors made another key discovery in the pursuit of a commercial light bulb. A long filament life depends on an oxygen-free environment. Light bulb pioneers employed a vacuum to protect the filament from burning up—if there is no atmosphere then there is no oxygen. Consequently, Edison owes much of his success to a near-perfect vacuum pump invented by the German chemist, Hermann Sprengel (1834–1906).

To reap the benefits of the complete package, a long-life electric lamp required a practical power source. So, in parallel with his light bulb, Edison developed plans for centralized electrical grids powered by steam-powered or hydroelectric generators. Ideally, they could supply steady electricity to clusters of homes and businesses, each with several electric lamps. The ambitious Edison planned to sell both electricity and light bulbs across the country.

On its journey to becoming a consumer staple, the light bulb not only depended on external help from power plants and electrical grids, it also needed internal improvements including reduced costs, better ambience, and longer life. Far from an instant success, it took years for the electric light industry to take hold. For an interesting perspective, consider the state of kerosene during the two-plus decades following Edison's famous invention. During that time, Standard Oil president, John D. Rockefeller, accumulated his fortune by refining and selling kerosene for lamp oil.

The first *modern* light bulb appeared in 1910, 31 years after Edison's patent. That year, General Electric (J.P. Morgan's gigantic version of Edison's lab) released its Mazda lamp. Its main advancement featured a tungsten filament instead of carbon. Thanks to a 1904 Hungarian patent granted to Alexander Just (1874-1937) and Franjo Hanaman (1878–1941), scientists knew that because of its superiority, a tungsten filament burned longer, brighter, and with reduced bulb blackening. Six years later, GE engineers broke down the high cost barrier with a practical process for manufacturing the tungsten filament. Engineered, drawn tungsten wire had a high melting point, high strength, and sufficient ductility. Manufacturers could make 75,000 40-watt bulbs from just one pound of tungsten. Another Mazda lamp innovation introduced the standard aluminum threaded base used today. Yet, more work remained to meet the light bulb's potential. And those innovations kept coming.

In 1913, lamp makers replaced the bulb's vacuum with low-pressure argon. The inert gas doubled luminosity and further reduced bulb blackening. A decade later, manufacturers refined the methods to frost glass. No trivial option, this innovation traded the harsh glare of clear glass with a proper glow, closer to the favored illuminating qualities of kerosene lamps. The chosen powder-coat method conquered three design obstacles. The frosting had to be inexpensive, noncorrosive, and minimize the blockage of light.

Light bulb makers continued to improve the incandescent lamp during the 1920s and 1930s resulting in a higher-quality and less expensive product. They increased the size of the coiling element and improved the recipe and purity of the inert gas. By the middle of the Great Depression, a modern incandescent light bulb produced ten times the light compared to earlier Edison models.

From Edison's time on, novel factory production methods occurred on a regular basis, both in smaller and larger increments. For example, as late as the early 1920s, artisans used hand-blowing techniques to manufacture glass bulbs. Whatever those production rates of the time were, they ran at a small fraction of the production rates of later twentieth century automated machines, which achieved a lightning-fast 50,000 bulbs per hour. Manufacturers repeated the cost-reduction process throughout the entire evolution of the light bulb, component by component.

The lighting industry invested countless hours to create the early twenty-first century incandescent light bulb, yet the physics of heating the filament to a temperature exceeding 5,000°F still results in huge inefficiencies. Less than 10% of the electrical energy consumed by an incandescent bulb converts to light, the remainder, wasted as heat. The cost of electricity equals several times the original cost of the light bulb. In response, the industry has turned its focus away from incandescent lighting in favor of alternative technologies such as compact fluorescent bulbs and light-emitting diodes (LEDs). Tomorrow, new technologies will emerge with a never-ending challenge to upgrade that next generation of inventions.

For two centuries, thousands of individuals added their input to form a massive matrix of light bulb innovation. A comprehensive study of the entire development of that technology might fill a small library wing. Edison contributed some larger chunks, and for that he deserves historical accolades, but while it sounds harsh, from an economic and technological perspective Mr. Edison is one of many.

In a process repeated countless times, most Great Depression and present-day products, whether consumer, commercial, military, or industrial follow their versions of development with their own complex histories.

## A Fresh Look at the Industrious 1930s

History books do not dwell on the pioneers producing 1930s technology. Instead they regard politicians standing up against the industrial machine the real heroes. However, in no previous decade did humankind achieve as much industrial enlightenment as they did in the 1930s—an endless flow of futuristic thinking. The combination of product design, manufacturing process, and material science drove a technological revolution that took industrialization to the next level. Consider the amazing feats accomplished by the advanced manufacturing technology of the Great Depression era. It launched the phenomenal military machine that achieved a victory against the powerful Axis powers. The decade's modern technology ended the Great Depression, and then provided a solid foundation for sustained, unequaled economic prosperity long after the war ended. Not a bad resume for what histography considers a lost-engineering era.

The following chapters feature dozens of major technologies, and although impressive on their own, appreciate that they represent a small percentage of the many fine examples of industrial progress from the 1930s.

# 33

# Fire the Chauffeur

Today, collectors seek 1930s automobiles as prized antiques, attracted to their elegant designs and bygone-era mechanics. Of course, car enthusiasts of 80-plus years ago viewed the new models as state-of-the art, an era in which automotive engineers elevated an automobile's utilitarian and social value. Cutting-edge technologies advanced the car's style, power, safety, reliability, and drivability—the latter the most important upgrade. The latest models offered unparalleled mobility available to an increasing portion of the population. Consequently, automobiles continued to modernize society—in the way people worked and their type of leisure—and all as they transformed America's landscape. The machine that exposed generations of teenagers to engineering and mechanics provides our first barometer of 1930s technological progress.

At the onset of the Great Depression, high-volume manufacturers produced the majority of automobiles. Even though the American industry broke an all-time sales record in 1929, with four-million new cars sold, automakers numbered less than 50 companies compared with 200-odd carmakers flooding the industry 10 years earlier—the economic reality of mass-production. By 1940, only 17 major auto manufacturers remained. The Great Depression contributed to the toll as American icons Cord, Pierce-Arrow, and Duesenberg met their demise, but there was more to the trend. Consider that during the prosperous 1950s, once-major brands Packard, Nash, and Hudson also went bankrupt. The movement hints that much of that decline reflected the competitive nature of a technological race, survival of the fittest in a world of mass production, and not the ongoing economic panic.

Whereas car manufacturers continued to focus on cost, art, and durability during the 1930s, engineers turned much of their attention to improving drivability, the automobiles of the day lacked the ease of operation, safety, and luxury features that drivers later became accustomed to. At the time, the skill level required to operate a car attracted the more adventurous, those with the proper skill sets and the dexterity required to negotiate a variety of controls.

Of course, all vehicles can be "unsafe at any speed" and treacherous at high speeds under any conditions. As a result, a significant portion of the population never learned how to drive. Some, if they could afford one, opted for an expensive option not supplied by the automakers—a chauffeur. It was a decision many regretted. Personal drivers gained a shady reputation because unlike other domestic help, chauffeurs spent too much time unsupervised. The potential freedom of an automobile created opportunities for mischief.

Innovative designs developed during the 1930s added incentives for a greater percentage of the population to learn the art of driving and for some to fire the family chauffeur. The new generation of user-friendly, family-oriented automobiles improved drivability and increased the comfort zone of the passenger compartment. They featured more trunk space, a smoother ride, and better heating and ventilation. Nash introduced air conditioning in 1939. Throughout the decade, cars got better gas mileage and burned less oil.

Automakers upgraded their non-synchro manual transmissions to synchronized gearing, which improved the ease of shifting through the gears. They improved the linkage and moved the shift lever from the floor, known as a wobble-stick, to the steering column, later referred to as three-in-the-tree. General Motors, after an intensive seven-year development program, introduced the first automatic transmission as a $59 option on select 1939 Oldsmobiles. Automatics became the scorn of automotive purists, but the delight of the public.

For a car's most important safety device, the brakes, manufacturers switched from mechanical to brake actuation by pressurized oil, called hydraulics. The leverage gained through the burgeoning field of hydraulics decreased stopping distances and improved stopping reliability. The Ford Motor Company, still influenced by Henry, became the final hydraulic brake convert in 1939. Other safety devices included safety glass for front and side windows, turn signals, greater visibility, side-impact beams, and reinforced, long-lasting tires.

More horsepower became yet another luxury for those seeking an entry level model—safety be dammed. Toward that goal, Ford introduced the first popular V-8 engine in 1932, a major automotive milestone. Their flathead V-8's combination of power, eventual reliability, and relatively low cost made it a preferred option that outsold Ford's standard-equipment four-cylinder engine five to one. V-8 Fords turned into the standard for gangster get-away cars and later launched the hotrod craze.

Consumers expected a lot from an automobile. On top of all the performance, drivability, utilitarian, and reliability requirements, a car had to have style. Consider the state of automotive design in the 1930 automobile model year. The top-selling but outdated Model T had finished its nineteen-year run just two years earlier in 1927. The next generation of models from top sellers Ford and Chevrolet, although larger and more powerful, had styling that remained boxy and gangly. They featured attached fenders and exposed radiators. Instead, buyers wanted modern-styled automobiles. Cars with a look that was more chic and less utilitarian. So, automakers abandoned styles with Victorian influences and turned toward an art deco motif.

As automobile design progressed through the 1930s, cars became more stylish, and per the fashion of the times, aerodynamic. Designers abandoned the harshness of vertical lines and sharp corners in favor of tapers, blends, and rounded edges. They modified the grills to make an artistic statement and moved them in front of the wheels. Automakers added an enclosed trunk, which integrated with the rest of the carriage. The automobile went from copying old art to creating new art, a prized makeover that did its share to overcome the recession's obstacles to the delight of the hard-to-please Great Depression consumer.

In 1934, General Motors' Fisher Body division revolutionized automobile body manufacturing. They combined larger sheets of steel, supplied by Inland Steel, with a newfound expertise in sheet metal-forming technology. Those advancements allowed automobile manufacturers to streamline and improve their designs making vehicles safer, better handling, and less expensive. For the conversion, Fisher Body used a series of four massive presses, each exerting 2,000 tons of force, to form an all-steel roof. GM dubbed their revolutionary body style the *turret top*. During the era of coach builders (pre-1934), automotive bodies lacked self-supporting rigidity. Instead, a car's exterior consisted of a wooden framework covered in canvas and sheathed in sheet steel. From the time of the turret top design on, builders configured auto bodies to be rigid, yet with a modern artistic flair. Structurally, the roof mimicked the keystone of an arch and included the front and rear window frames. Along with spot welding, the new design reduced vibration allowing for more power and higher speeds.

1935 Chevy Master Deluxe Turret Top. *Courtesy GM Media Archives.*

Advanced automobile designs and the ever-expanding network of roads inspired the iconic American pastime—travel trailers. By 1940, Americans owned nearly one-million sleeper trailers.

Yet even with all those improvements carmakers maintained steady or declining pricing. The cost-busting strategy automakers mostly relied on was economies of scale. The best example worldwide for any industry was Ford's River Rouge complex completed in 1940. With an output of 6,000 automobiles per day (one every ten seconds) it was the largest factory in the world. The scale of the industrial city, which supported 80,000 workers, reached an unimaginable size. The logistics alone required to handle the enormous amount of parts and raw materials boggles the mind. Thus, despite the Feds relaxing tight monetary standards, Ford and Chevy kept the cost of a new car well under $1,000 throughout the 1930s, which was less than the price of their 1930s entry-level models. To round out the Big Three and with a longer-term comparison, Chrysler's entry-level 1919 Dodge sold for $2,160—a 1936 for $800.

1939 Ford DeLuxe convertible coupe, *Courtesy The Henry Ford.*

Automobile sales leveled off during the 1930s, yet automakers sold an impressive 25 million models during the decade and another 7 million in 1940 and 1941 combined. For any economist suggesting a drop in productivity are not considering that any throughput shortfalls were more than offset by significant gains in added value. In other words, as the market struggled to put Americans back to work during the Great Depression, the nation's leading industry joined many others by increasing productivity during the Great Depression.

Transportation to fair - Women with car and trailer. *Courtesy Manuscript and Archives Division, New York Public Library.*

# 34
# From Pickups to the Big Rigs

A bustling industrial nation depends on an advanced commercial transportation system. Railroads' ability to haul heavy loads over long distances made them one of the key contributors to the Industrial Revolution. After 30 years of development, the 1930s trucking industry made their dramatic impact as they filled a logistics gap left unfilled by railroads and horse-drawn wagons, and only partially satisfied by the performance of pre-depression trucking. The versatile truck, with great efficiency, brought raw materials to a factory's doorstep and delivered finished goods to distributors and retailers.

Truck and bus makers throughout the 1930s made impressive gains. The differences are highly visible. Whereas trucks from the 1920s have that old-fashioned look, trucks from the 1930s feature relatively contemporary designs. Despite the economic slump, over a dozen manufacturers maintained a vibrant industry. The tractor-semitrailer, diesel engine, streamlined bus, and cab-over-engine design all became future standards during that decade. Of those, Ford, General Motors, Dodge (later re-badged as Ram), Kenworth, Peterbilt, and Mack still make trucks today.

Similar to the railroads a century earlier, the trucking revolution altered the country's ability to move goods and along with the automobile, reshape the American landscape. First, surrendering to stiff competition from trucking, the teamster and his team of majestic draft horses disappeared. Next, diesel buses, with the help of automobiles, stole a significant number of customers from the once highly lucrative business of electric trolleys and passenger trains. Because of their huge capacity, mighty freight trains survived the heavy competition. Each with their own advantages, freight railways and trucks shared long-distance hauling, but the

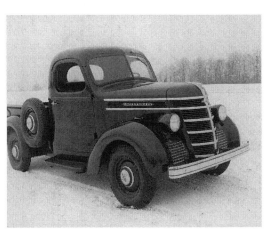

International 1939 D-Line Truck. *Courtesy Wisconsin Historical Society.*

practicality of trucks for local deliveries attracted enough additional commerce that both industries continued to grow after the recovery.

The most famous names in the trucking industry transformed during the 1930s. Mack, the former wagon maker, manufactured a full range of trucks, including pickup trucks, buses, medium-sized trucks, fire trucks, and cross-country rigs. White, Kenworth, and Peterbilt made heavy-duty trucks and fire engines. Diamond T showcased robust designs of various sizes. Another icon of the era, REO, might be more familiar to today's readers by the name of its most popular model, the Speedwagon. International Harvester moved from farm equipment to a vehicle range that spanned from light to heavy duty. The big three, General Motors, Ford, and Chrysler, all competed in the truck market. The dominant automakers offered a full range of trucks, although they became most famous for their pickups, an essential for the American farmer and tradesman.

One of the telling differences separating the older-styled, post-World War I trucks and buses from their modern successors are the wheels. Once again, humankind reinvented the symbol of civilization. By the onset of the Great Depression, trucks made the conversion from spokes to heavy cast iron spoked wheels and hubs. By the early 1930s, the trucking industry revolutionized the wheel, standardizing on the modern steel two-piece disk design. The advantages of modern wheels included higher load capacity and the ability to remain in alignment despite collisions with road hazards.

In a classic case of new technologies exposing the weaknesses of another, tire manufacturers needed to design tires to handle heavier loads and higher speeds, while improving reliability. Truck tire leaders such as Goodyear and Firestone met those requirements with heavy-duty pneumatic tires. Innovations of the era included more durable synthetic rubbers and corded reinforcements.

For faster and more reliable braking, trucks, as did cars, upgraded their mechanical braking systems. Smaller trucks adapted automobile-styled hydraulic brakes. Heavier trucks used air brakes, which had over

Semi-tractor with 5th wheel-1939. *Courtesy Farm Security Administration – Office of War Information*

Pre-war Kenworth Fire Truck. *MOHAI, Courtesy Museum of History and Industry.*

International COE Coca-Cola truck-1938. *Courtesy Museum of Industry and Science*

50 years of proven experience on trains.

Bigger trucks employed larger engines for two reasons—to provide more power and to reduce the strain of heavy-duty cycles. Truck manufacturers addressed the issue through a couple of different paths. For the first option, they increased the size of their standard four-cylinder gasoline engine by adding cylinders. General Motors and most others went to an inline-six configuration. Ford offered a truck version of its famous automobile V-8. To give those engines even greater versatility under varying conditions and deliver either high torque or high speed to the wheels, depending on conditions, truck makers used four-speed transmissions on smaller trucks and seven-speed transmissions on larger trucks. The greater revolution to trucking came when truck builders commercialized the diesel engine.

The development of the small diesel engine revolutionized the trucking industry. Mechanically similar to its gasoline-fueled counterpart, Diesel's engine operate on a principal of compression ignition, as opposed to gasoline engines, which operate on spark ignition. Diesels fit the heavy-duty requirements of trucking perfectly. Compared to gasoline versions, diesel engines have higher efficiencies, operate at lower rotation speed (RPM), have greater torque, better lubricating qualities, require a less-expensive fuel that packed more energy, and are designed for heavier duty cycles. All those features meant that diesel engines for an equivalent horsepower delivered better fuel economy, nearly double that of gasoline, with impressive durability.

German engineer Dr. Rudolf Diesel patented his namesake innovative version of an internal combustion engine in 1892. Rudolf Diesel's unexplained death in 1913 on a transatlantic cruise, which stands as one of the twentieth century's great unsolved mysteries, left others to promote his engine design. By the 1920s, submarines, along with a few surface ships, were the only vehicles with practical and profitable applications for diesel engines. The early years of the Great Depression had several manufacturers closing in on diesel engines for land transportation. Ingersoll Rand had experimented with a diesel locomotive engine since 1917, and from the early 1930s, Caterpillar developed a diesel engine for off-road tractors, and Cummins for trucks and buses.

Beginning in the 1920s, Clessie Cummins (1888–1968) built and sold British-licensed diesel engines for stationary applications, mostly to farmers. He later expanded his market to include marine applications. In 1932, the Cummins Engine Company introduced their own version of a working diesel engine for motor vehicles. To promote the novel alternative to gasoline engines, he came up with a series of remarkable public relation stunts. In his most famous exploit, Cummins installed his diesel engine in a Packard for an attempt at the world nonstop automotive distance record. He made his challenge in 1933 on the Indianapolis speedway. The Cummins-Packard completed 3,000 laps, 14,000 miles, without a single pit stop (he did refuel while moving), an incredible feat for a startup engine builder. The following year, Cummins was in business building diesel engines for trucks and buses. Despite his successes, over the next four years he struggled to earn a profit, so Cummins made a bold and risky decision. In 1937, his company included an unprecedented 100,000-mile warranty with the purchase of a Cummins diesel engine. From that time on, despite being in the depths of the Great Depression, the Cummins Engine Company became profitable. Later, General Motors, Ford, and others joined the diesel manufacturing market. There was room for all, as diesel engines replaced the bulk of gasoline engines across the trucking and busing industries.

For the 1930s, truck manufacturers made other dramatic modifications to a wide range of vehicles as they became more modern in performance, form, and function. One of the universal changes truck builders made on medium-duty and heavy-duty trucks is that they moved the engine from behind the front axle, similar to automobiles, to over, or even in front of it. The new weight distribution improved load capacity and handling. From there on, the challenge became: get the most cargo space into the shortest vehicle, while maintaining high safety standards.

Semi-tractors improved the task of transporting goods over longer distances. The term semi-trailer comes from the way a trailer couples to a tractor's fifth wheel (see photo below), which eliminates the need for the trailer's front wheels. The fifth wheel is so robust that even today it is the only trailer coupling that does not require a safety chain. These articulated tractor-style designs, also known as *road trains*, make a lot of sense. They are an elegant solution to moving freight, which is why their popularity continues. Instead of offering a wide variety of specific-purpose trucks, manufacturers built only a handful of basic tractor models. The responsibility of specialization fell on the trailer builders. Only size and load restrictions limited trailer designs, otherwise designers could push the boundaries of their imagination. The logistics of having the flexibility to separate the trailer from the truck offered other major advantages for a variety of industries. For example, the storage, loading, and unloading of a trailer did not have to waste the truck's or the driver's time.

In 1933, Kenworth installed Cummins diesels in their heavy-duty trucks. The diesel's characteristics of great fuel economy and durability showed promise for the then untapped venture of cross-country freight hauling. Besides a diesel engine, for longer trips, Kenworth designed a larger cab that incorporated a sleeper compartment. With the other upgrades including longer-lasting tires and improved highways, such as Route 66 in the west and Route 1 along the east coast, diesel tractors established long-distance trucking.

For in-town maneuvering and parking, truckers needed to pack maximum cargo into the shortest truck. To that aim, manufacturers introduced the cab-over-engine (COE) truck design in the early 1930s. Legal weight limits for roads were per axle. The cab-over design evenly spread the load allowing a 75% increase in hauling capacity compared to traditional trucks. The basic configuration of the driver sitting above the wheels and beside the engine remains similar to this day.

Before the Great Depression, companies such as Mack built buses from modified trucks; however, the designs originating from the 1930s set the standard for buses used today. Modern highlights included the driver sitting ahead of the front wheels, a diesel engine located behind the rear wheels, a step-up entrance across from the driver, storage between the wheels, the passenger compartment above the wheels, and overhead storage. Passenger comforts included lighting and ventilation. For driver upgrades buses had air brakes and pneumatic controls to help shift the transmission. The coach-like, streamlined body had a wide track, lowered chassis, and a liberal use of aluminum. Buses became so practical for mass transit that after 1930 few passenger trains would ever again have a profitable year. By 1938, 130,000 buses operated in the United States. Buses offered greater flexibility compared to either trains or trolleys, which matched the transportation needs from congested urban areas to suburban sprawl.

One more iconic truck worth mentioning that arose from the 1930s is the Divco delivery truck. If the reader does not know the name, then they should recognize the vehicle shown. In 1937, Divco introduced what would become America's standard milk-delivery truck. Delivery vans were designed for frequent stop and go which required the delivery man to make a quick exit and entrance. The Divco featured a welded, all-steel body forming a single compartment, a folding side door, and an economical four-cylinder engine. Despite the frequent exchange of companies in charge of building Divco vans for decades, the

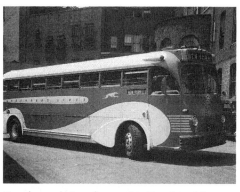

Greyhound Lines bus-1937. *Courtesy Library of Congress.*

Divco delivery truck stands as one of the longest-running motor vehicle models ever built. From the late 1930s into the 1980s, the classic Divco snub-nose trucks served as special purpose multi-stop delivery vans that not only delivered milk, to compliment the rise of home refrigeration, but they transported many other household and commercial staples from bakery goods to laundry.

Compare the stark contrast of the 1920s truck to the subsequent images of the 1930-model trucks in this chapter. Beyond their radical aesthetic upgrades, appreciate that a considerable amount of technology churns beneath those sleek exteriors. Critical attributes such as speed, power, safety, hauling capacity, and reliability all improved during the era. To add to the productivity equation, and despite the recession, truck makers sold more vehicles during much of the 1930s compared to the late 1920s. If someone unfamiliar with the history of the Great Depression first looked at those photos, they might be surprised to learn that those dramatic enhancements occurred during those dismal times.

Milkman in front of Divco delivery truck—1938. *Courtesy Museum of History and Industry.*

To grasp the full impact of trucking technology on the economy, consider the scope of the market trucks served. At some point, the transportation for all the nation's goods required a truck, and often on multiple occasions. The progression that trucks achieved during the Great Depression played a significant role toward the eventual robust economic recovery. The evolution

A driver sits in an International truck TiltTop Bottle Capper-1929. Courtesy Wisconsin Historical Society.

of trucking is another example to make one wonder why historians ignored the technology produced during that time.

A final note regarding motor vehicles—automobiles and trucks. Americans traveling the nation's highways consumed 14.1 billion gallons of fuel in 1929, double the consumption of five years earlier. Despite the hard times, US fuel consumption continued to rise during the Great Depression reaching 20.7 billion gallons in 1939.

# 35

# The Commercialization of Flight

Airplane design underwent dramatic advances during both the 1920s and the 1930s. For the 1930s, flight manufacturers reached a historic milestone—the widespread growth of commercial aviation. The prodigious amount of engineering required to make flying safe and profitable also strengthened the industry's reputation as one of the most prestigious technologies.

At the beginning of the decade, the majority of aircraft were biplanes; 10 years later single-winged, high-tech planes dominated the skies. The dramatic success of a single model, the Douglas DC-3, highlights the remarkable technological progress gained in the field of aerospace during the Great Depression.

Aircraft safety engineering progressed throughout the 1920s. That sense of urgency intensified with the well-publicized crash of TWA flight 599, known as the Rockne Crash, named for Notre Dame's legendary football coach Knute Rockne. On March 31, 1931, a Fokker carrying Knute Rockne from Indiana to California suffered a catastrophic mechanical failure. The plane lost a wing while in flight and crashed into a Kansas field. The culprit blamed for the fatal crash was structural integrity. The Fokker trimotor, unlike the all-metal Ford, used laminated wood for part of the structure. TWA favored the lighter Fokker for its higher performance. Predictably, the wood turned out to be the source of the failure. From then on aviation safety experts determined that wood had to go. Wood was susceptible to defects which were difficult to detect, and rot could not be welded or riveted. The aircraft industry did not care for the Ford Trimotor design either. Trimotors were boxy, the wing supports took up valuable cabin space, and excessive cabin noise, vibration, and fumes were not passenger friendly. The airline industry decided they needed to take a giant leap forward.

Boeing was the first to build the next generation of radically redesigned airliners. In 1933, Boeing introduced their Model 247 airliner with its commercial partner United Airlines. The revolutionary 247 made huge advancements in aviation. Some of its design features included an all-aluminum structure, two 700 horsepower engines, cantilevered wings, variable-pitch propellers, and retractable landing gear. The 247 also had the capability to fly with one engine. (The Ford Trimotor lacked the critical safety feature of being able to maintain flight with two of its three motors operational). At 200 mph, the sleek and powerful Boeing flew faster than any commercial aircraft.

Transcontinental and Western Air (TWA) requested an order for the 247 aircraft, but Boeing denied the offer. Boeing had allocated its first production run to its partner, United. TWA then turned to the Douglas Aircraft Company and with funding from the airline, Douglas developed their version of a modern aircraft, the DC-1. In July 1933, Douglas successfully tested a DC-1 prototype. A year later they released the DC-2. A moderate success compared to the Boeing 247, the DC-2 had a more modern, aerodynamic design. Douglas sold 193 DC-2 airplanes.

Impressed with the DC-2, American Airlines approached Douglas with a proposal to build an airplane with sleeper cabins instead of traditional seating. Douglas accepted the order but did not expect high sales volumes because at the time the public avoided nighttime flying. American requested the plane be larger and that Douglas upgrade the plane's directional stability. Douglas believed they could use 90% of a DC-2, but with the design completed they realized that they borrowed only 10%, so they renamed their latest design the DC-3. The DC-3 flew its inaugural test flight on December 17, 1935.

In 1937, the Douglas Aircraft Company, backed by American Airlines, formally introduced its iconic DC-3 airliner, known as the Gooney Bird—one of the most legendary planes of all time. It had so much early success that it allowed commercial aviation to reach that historic milestone—profitable passenger service without generous subsidies from the US Postal Service. Douglas planned to sell 100 of them—by 1940 that number approached 1000. More impressive is that in that same year, DC-3s accounted for 95% of the nation's passenger miles.

Douglas built a remarkable bird. With its long, monocoque fuselage, powerful enclosed twin engines, and swept-back wings, the DC-3 looked remarkably modern. Each engine produced 1,200 horsepower (2,400 HP total), which allowed the DC-3 to carry up to 34 passengers and still cruise at more than 200 miles per hour. It had a cruising range of about 1,500 miles, twice that of a Boeing 247. In other specification comparisons, a single 110 horsepower engine powered the World War I British fighter plane the Sopwith Camel 20 years earlier. Ford Trimotor airplanes of the 1920s came equipped with three engines each with an output of up to 400 horsepower. TWA's Fokker cruised at less than 100 miles per hour.

Tailwind transcontinental flights on a DC-3, west coast to east coast, took about 12 hours. The return flight lasted two hours longer due to the additional drag of headwinds. In the accompanied photo notice the DC-3's signature stance, earning the planes of that era the moniker *tail draggers*.

Proof of Gooney bird's great success lies in the number of DC-3s produced. Douglass, combined with its foreign licensees, manufactured over 17,000 DC-3s, including its military variant, the C-47. Production of the great bird ended in 1944. A vast surplus of con-

Elevated view of group of men and women boarding an American Airlines airplane while an International D-15 truck marked "U.S. Mail" and "American Airlines" is parked in the foreground. *Courtesy Wisconsin Historical Society*

verted WWII military models found a second life as civilian aircraft after the war. Its utilitarian value continued as the DC-3 served every major airline for decades. Incredibly, a handful of those vintage planes flew commercial routes into the twenty-first century.

Pan American World Airways (Pan Am) created a niche market with a fleet of spectacular flying boats known as clippers. Both Martin and Boeing manufactured the big birds. With pontoons in place of wheels, they did not depend on airports, but instead landed in protected bays. Not as commercially successful as the DC-3, but they were still historically significant for their many *firsts* in long-distance commercial flights.

The massive clippers mostly came in sleeper versions and of course with a high-ticket price. The airline targeted a high-end market and so no surprise that their clients included celebrities and world leaders. The most famous aircraft in the fleet was the China Clipper first flown in 1935. A 1936 film entitled the *China Clipper* starred film legend Humphrey Bogart.

Powered by four 1,500 horsepower Wright engines, those 82,500-pound flying birds featured two decks. There were approximately 18 luxurious passenger cabins, plus bathrooms, a galley, and a dining room.

With a 4,000 mile-plus range they specialized in long transoceanic flights, which included scheduled routes over both the Atlantic and Pacific oceans.

Aerospace's need for light weight, efficiency, durability, safety, and high precision led the industry to function as an engineering hub with their craft relying less on black art (empirical knowledge) and more on pure science. That technical knowledge then transferred to a slew of other industries.

The proof of their engineering proficiency showed in the speed plane builders moved from concept, to prototype, to production. Manufacturers delivered safe, well-built machines in less than two years. The partnerships between aircraft manufacturers and airlines produced those impressive results at a time of limited research and development funding.

By the end of the 1930s, the United States counted 2,500 airports, including seaplane harbors. The large network incorporated wireless radio technology, which added several layers of safety with pilot-to-pilot and ground-to-pilot communications. Those are impressive accomplishments for an industry that did not exist at the beginning of the century, and therefore, once again, the history of technology offers yet another contradictory insight into some vigorous commerce overshadowed by the despair of the Great Depression.

# 36

# The End of the Cracker Barrel

Modern American families turned to commercial food products as locally grown food, both from farmers' markets and backyard gardens gradually, gave way to the grocery store. Technology-driven, store-bought food goods depended on a vast and complex network of farms, processing plants, haulers, and distributors. The latter then delivered the merchandise to local markets, chain grocery stores, or doorsteps. For those products to be practical in high volumes, they had to have affordable and reliable packaging. In response, the packaging trade developed into a high-technology industry. Similar containers also served other markets such as cosmetics and pharmaceuticals. Three of the largest packaging industries of the Great Depression were canning, bottling, and box making.

## Canning

The history of canned foods dates to the early nineteenth century, originating in France, the England, and the United States. Returning soldiers brought their taste for canned foods home creating demand for the civilian market. New technological breakthroughs during World War I led to the canning industry's first large expansion—the accumulation of more than 100 years of canning knowhow. But canners had work to do. The twentieth century soldier's changing tastes demanded a higher quality cuisine.

Over the next 15 years, foodstuff canning expanded, buoyed by increasing productivity and a wider variety of food choices. The industry accomplished this feat through the modernization of the three primary phases of canning: food preparation, can manufacturing, and finally, the canning process.

Food preparation science first modernized in the 1890s. By that

Empty cans are filled with peas and lids placed on them. Canning factory, Sun Prairie, Wisconsin -1937. *Courtesy Farm Security Administration-Office of War Information*

time scientists from universities, trade organizations, and government agencies had perfected the practical methods needed to eliminate contamination by harmful microorganisms. The standard practices they developed included the use of either heat, cold, or dehydration (freeze-drying). Researchers also gained knowledge on how to maintain the product's nutritional value and good taste without excessive preservatives. Later, in the early twentieth century, the widespread use of stainless steels throughout the numerous preparation processes improved the hygienic practices of the industry.

The modern tin can originated in Europe during World War I. The scale of the Great War made the traditional, labor-intensive and inconsistent soldered can impractical. The replacement was similar to the can we are familiar with today, a welded barrel capped with a stamped, double-seamed lid and bottom. In its modern form, automatic welders and presses had replaced the craftsman.

The formed and welded can proved to be an elegant design. For the material, canners used sheet steel for strength, which was coated with a thin layer of tin to prevent rust. The double seam provided a near-perfect hermetic seal, giving the can its extended shelf life.

Approximately 3,000 canneries operated in 46 of the 48 states. They packaged over 300 varieties of fruits, vegetables, meats, seafood, and dairy products. Manufacturers located their canneries near farms or seaports to reduce transportation costs and maintain freshness. For precise uniformity, each farm raised or grew their products to strict standards directed by the canneries. For example, the carrots for a brand of canned carrots would have all been grown from the same variety of seeds.

The technology available through electric factories became the basis for canneries to create highly automated methods, which combined food processing expertise with high-speed canning machinery. Speed and consistency were critical for both food safety and low costs. By the mid-1930s, on average, each automated line produced 300 vacuum-sealed cans per minute compared to a turn-of-the-twentieth-century artisan which produced one can per minute. Annual sales volumes of the more popular canned goods, such as pineapples and condensed milk, ranged from 50 million to a few hundred million. By the late-1930s, automatic machines produced nearly 10 billion cans per year.

The canning industry exemplified a major economic sector turning modern during the Great Depression. The complete package involved special harvesting equipment dedicated to various fruits or vegetables, an expanding system of roads and heavy-duty trucks to bring the raw materials in and the finished goods out, and sophisticated machinery that counted on numerous trades within the larger canning industry.

Canned foods were well suited for the growing number of urban families. Even more important, their low cost and high nutritional value served well during the tough times of the Great Depression.

## Bottling

The bottling industry followed an early commercial path similar to canning as manufacturers developed much of the base technology in the late nineteenth century. Prime candidates for bottling included foodstuffs, carbonated beverages, alcohol, milk, and medicine.

Bottle-making reached a high degree of mechanization after the turn of the century when in 1903 Michael Owens (1859-1923) established the Owens Bottle Machine Company out of Toledo, Ohio. His top-performing, massive mechanical marvels weighing up to 18,000 pounds automated the entire bottle-making process. By the 1920s, Owens machines could produce a remarkable 4 bottles per second. The machines were also flexible enough that with a tooling changeover they could produce a wide variety of bottles. In 1929, the renamed Owens Bottle Company merged and became the Owens-Illinois Glass Company.

To be viable, bottle making had to be inexpensive, requiring the streamlining of the entire process—from the ingredients to the final operations. In the 1930s, the primary raw materials used for container glass were soda ash, limestone, and silica sand. Additional ingredients improved transparency, or gave the glass a colored tint such as amber or green. Standard glass recipes included recycled broken glass (cullet), which augmented the melting process. In the melting process, a furnace produced the gob, a piece of glass made to precise dimensions.

A ten-arm Owens automatic bottle machine, ca. 1913. *Courtesy Lewis Hine*

The molten gob then passed through various complex processes. The primary steps were mechanical glass blowers, which used compressed air and molds to shape the gob into the rough form of a bottle. Additional stages used either a vacuum or mechanical means to bring the molten glass to its final shape. The bottle then had to be rapidly cooled to maintain the glass's integrity. The bottle was then warmed in an annealing furnace, called a lehr. The lehr increased the glass's durability. For the complete automated package, a network

Filling milk bottles at creamery. San Angelo, Texas - 1939. *Courtesy Farm Security Administration-Office of War Information*

of motorized conveyors ran between the bottle-making operations.

Some bottle designs of the 1930s became iconic. The Coke bottle gained international fame. Heinz, instead of style, trimmed their ketchup bottle to fit conveniently into another modern 1930s high-technology product, the refrigerator.

By 1939, consumers purchased nearly two-and-one-half billion jars of food and over two billion beverage bottles each year. The average cost per bottle was less than three cents. Those volumes represented a seven-fold increase from 1900, yet high-tech automation resulted in a 10% reduction in the number of bottle-making workers.

## Box-Making

Box making, along with cartons (a smaller version of a box), developed into another significant packaging industry during the Great Depression. As with other packaging industries, practical box making was also a highly engineered process. Its long line of construction steps began with the continuous processes of paper making followed by the paper-converting machines, which produced either plain or corrugated fiber board. The box industry shuns the term cardboard. The key to high production was the box-cutting machines. High-speed hydraulic rams stroked precision dies that simultaneously made cuts and score marks. Depending on the type of box, finish operations may have included printing, folding, and gluing. For heavy-duty applications, box makers added metal edging. High-speed machinery allowed the nearly 60,000 workers in the American box making industry to produce a prodigious quantity of boxes.

Placing packaged goods on display rack. Retail grocery, San Angelo, Texas - 1939. *Courtesy Farm Security Administration-Office of War Information*

Most consumer products before the 1920s shipped to the store in bulk, but with commercialization during the 1920s and beyond, manufacturers packaged their products in customized boxes. Over time, the cracker box replaced the cracker barrel. Paper packaging kept their products clean, dry, and germ free, but maybe most important, the box could promote their brand. The movement from bulk to box also started the transition from commodity to trademark, another trend of the modern-consumer society.

Cans, labels, and bottles featured embossed logos to promote the brand names, but boxes had multiple flat surfaces—more opportunity to catch a potential customer's eye. Those facades provided a durable surface for informative text and dynamic graphics. For the advertising promoter, the box's exterior became a venue for the artist.

## Packaging Conclusion

For the student of economics, it is important to note how much the modernization of the not-so-glamorous packaging industry added to the nation's wealth during the Great Depression. At first glance what appears to be a mundane product blossoms into a universe of technological marvels that boosted national wealth.

The means by which the 1930s packaging industries modernized and benefited the economy was counterintuitive. Even though the packaging industry yielded double-digit annual growth rates during the 1930s, they accomplished that expansion with few additional new jobs. Instead, those companies relied on mechanization to increase their corner of the economy. Those inherent benefits ran counter to conventional wisdom's New Deal goals of creating many jobs no matter how frivolous the outcome of that labor. The elegance of labor conservation is that it directed surplus workers toward other economic sectors. As a result, the American economy grew by squeezing in additional industries—the true source of economic growth.

# 37

# Larger, Smarter, and More Versatile Machine Tools

The advancement of machine tools, credited in Chapter 8 as one of the key pieces of equipment responsible for the Industrial Revolution, continued to progress throughout the 1930s. The post-1920s era marked a transition period for the industry, as a new generation of machines switched from line-shaft pulleys to efficient electric motors. Electrification also encouraged an array of accessory gadgets that increased their speed, versatility, and reliability.

The Great Depression hit capital industries hard, cutting manufacturing sales revenues on average in half. Fortunately, America's world-leading machine tool industry, despite plunging revenues, kept busy throughout the 1930s, advancing their designs in preparation for the eventual recovery. One source of income came from the Soviet Union. The fledging communist nation purchased lots of machine tools for their ongoing Five-Year plans. With the industry in a lull, the ruble furnished valuable funds to finance research.

Once a purely mechanical industry, by the 1930s machine tools benefited from a multitude of disciplines. Technological upgrades included those from the growing field of electrical engineering, which provided machine tools with high-efficiency electric motors, along with a number of specific-duty sensors and switches. Engineers also learned to control machines with either compressed air (pneumatics), or by pressurized oil (hydraulics), a field led by the aircraft industry. Machine tool builders further took advantage of another emerging fluid discipline, centralized lubrication, which allowed spindles to rotate faster and increased the life of all moving parts. To boost cutting speeds, machine processes began to rely heavily on machining coolants (liquids that both lubricated and chilled the work zone). The steady stream of technological advances added speed, power, and precision to industry's machines that make machines.

Besides engaging in complex technologies, machine builders seized the advantages of larger equipment; for example, the multi-story hydraulic presses capable of producing 50 tons of force used by the aircraft and automotive industries. Machine builders went big for two reasons. First, their customers were building larger pieces of factory and mill equipment, i.e.: tractors, airplanes, locomotives, ships, electric turbines. Bigger machines required bigger machine tools. A second reason was to increase profitability by taking advantage of economies of scale. This was the same strategy pursued by Leonardo da Vinci, the famous part-time painter and sculptor, full-time industrial and military equipment designer.

Another challenge for machine builders was their customer's use of more complex shapes, which required more sophisticated machines. One of those groups of machines, introduced in the 1930s and designed to meet those demands were die-sinkers. Die sinkers utilized all the latest motor and sensor technology of the day, plus revolutionary hydraulic controls to produce intricate forms.

The introduction of new types of grinding machines modernized the industry. Internal grinding and jig grinding machines employed high-speed spindles exceeding 50,000 RPM, which allowed machinists to produce more accurate holes at faster speeds.

Superfinishing was another technology reaching maturity during the 1930s. Using a stone similar to a precision grind wheel, but with smaller grains, a superfinisher moved minute amounts of material to produce a near mirror surface finish with greater uniformity, just another technology that squeezed greater performance and longer life out of an internal combustion engine.

Driven by high production and lower costs, the automotive industry was the leader pushing more productive machines. One such machine tool was a motor-driven, gang-styled drilling machine that simultaneously drilled one-hundred holes in an engine crankcase.

## The Bridgeport Mill

One of the most notable and innovative machine tools to emerge out of the decade was the Bridgeport milling machine, usually just referred to as a Bridgeport. The invention of a Swedish immigrant, Rudolf Barrow (1903–1972), the Bridgeport became the most iconic machine tool in American history. Barrow did not invent the milling machine—mills date back to the early nineteenth century. However, Barrow packed his universal edition with lots of versatility, incorporating features not available on traditional mills.

Introduced in 1937, and named after the Connecticut city in which it was manufactured, the Bridgeport mill incorporated loads of pioneering technology. It featured a ½ horsepower electric motor with a rotating ball-bearing-supported spindle. Power from the motor to the spindle transferred through a variable speed drive consisting of stepped sheaves, along with a recent innovation called a V-belt. V-belts, made from synthetic rubber or plastic and named after their shape, transmitted higher torques within a smaller space than the traditional leather flat belts of the time.

Flexible machining requires a wide variety of tools, which for efficiency need to operate at various speeds. The Bridgeport design allowed a machine operator, through a system of levers, hinges, and latches, to change spindle speeds, or the direction of rotation, by the press of a button. An ingenious quill-type slide allowed for either rapid movement or precise adjustment of the cutting tool. The clever, precision quick-change collet held a wide variety of tools.

To handle an assortment of shapes and sizes, a compound-dovetail slide arrangement with generous travel moved the work laterally under the cutting tool. Each slide had its own individual precision lead screws terminated through a hand crank with a graduated scale. For the rough vertical adjustment, the Bridgeport came equipped with a large dovetail and another hand crank located near the machine's column. The worktable had T-slots for the easy mounting of clamping devices.

The Bridgeport milling machine complimented another shop staple, the metalworking engine lathe. Whereas a lathe profiled rotating parts. Barrow's universal mill, fabricated parts of various non-cylindrical shapes with almost as much ease.

**Figure 37.1** Turned part from lathe

By 1930, the turning of cylindrical parts had reached a high level of sophistication. A machinist with just a single lathe, by changing out cutting tools, could produce the complex shaft shown in Figure 37.1, which includes making the barrel shape (turning), smoothing the rough edges, adding chamfers and fillets, and then finally cutting the external screw thread.

Before the introduction of the Bridgeport mill, the part shown in Figure 37.2, of approximately the same complexity as the one in Figure 37.1, would take about three machines to fabricate, depending upon the shop. Machining operations would include the facing of six sides, making a slot, chamfers, plus

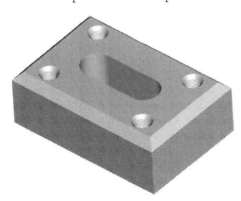

**Figure 37.2** Milled part from Bridgeport mill

drilling, and tapping four threaded holes for machine screws. In the early 1930s, the time it would take to produce the part included the machining time, wasted time for proper scheduling of those three machines, time to transport the part, and time to realign the part in each separate fixture. A Bridgeport was the only machine required, along with several rapid changes of tooling—a much quicker and more accurate solution to part making.

The Bridgeport mill's unique, general-purpose design and limited robustness did not suit high-production manufacturing; instead, the bulk of them served machine shops. The machine shop is one of the great, unsung enablers of any healthy industrial economy. Equipped with an assortment of machine tools and operated by specialized technicians known as toolmakers, machine shops fashion or modify components for machines and products. Even during the Great Depression, the American economy required thousands of those workshops. Many were small: one or two toolmakers equipped with a handful of basic machines. They might support local farmers, neighboring factories, or when based on board a ship, emergency repairs. On the other end of the spectrum, large corporate research facilities would have employed over 100 toolmakers in their prototype labs. Each of those well-furnished shop's repertoires included hundreds of machines of different sizes and each built for special purposes.

The launching of America's technological wonders, large and small, such as automobiles, airplanes, home appliances, and radios, along with thousands of other items, relied heavily on machine shops. When designs left the engineering department, they went directly to the toolmakers for prototyping. On the production side, most of the manufacturing processes within a factory depended on machine shops. For example, toolmakers fabricated the parts for the many special machines, tooling, and fixtures needed to make, inspect, package, and transport factory goods.

For America's heavily industrialized economy, the Bridgeport mill was a game-changing invention, influencing the production of many goods, often numerous times within a product's string of manufacturing processes. Because of its versatility, the Bridgeport mill made the manufacturing workshop, with its growing repertoire of machine tools, more productive, thus increasing the proficiency of virtually every industry.

By the late 1940s, most machine shop, from the small one-man operation to the mega high-technology shops supporting the largest factories, had at least one Bridgeport. Over the next 50 years, sales estimates place the volume of universal milling machines built by Bridgeport at more than 250,000. That is a huge number of machines for an industry that purchases a machine with a useful life covering decades and rarely scraps such a versatile and valuable machine.

When competition arrived from similar universal milling machines, the industry considered their mills a generic product, referring to those clones as Bridgeports.

# 38

# Super Alloys

Metals formed the foundations of advancing civilizations. The most influential had ages named after metals: The Copper Age, Bronze Age, and Iron Age. Bronzes and irons were not only strong, but their unique characteristics allowed craftsmen to make tools into complex shapes, which made the metals valuable for many applications. To propel America's mid nineteenth-century industrial revolution required a new generation of game-changing materials capable of taking technologies into the next age.

Materials Advancements Pre-1930

Over its first 100 years, American industry relied on age-old standard materials such as wood, iron, copper, brass, and bronze. They formed the backbone of the Industrial Revolution. From the Gilded Age onward steel played a critical role in the continuing industrial revolution.

The most advanced material in America's early industrial repertoire was steel, an expensive hyper-version of iron. As with cast iron, steel is extremely strong in compression, but unlike traditional iron, steel is capable of high bending strength. The latter characteristic significantly increases steel's utilitarian value.

For centuries civilizations produced steel most notably in Syria, England, Sweden, and Japan. While each of those societies produced high-quality steel, their manufacturing costs remained high—consequently they produced limited quantities.

America has an abundance of steel's main ingredients (iron ore, lime, and carbon), but even after the Civil War the nation still lacked practical methods to mass-produce steel. In 1875 Scottish-born ex-railroad executive Andrew Carnegie opened the Edgar Thomson Steel Works plant in Pittsburgh, Pennsylvania. The mill used a revolutionary steel-making process named for its English inventor Henry Bessemer. The Bessemer process produced high-grade steel in large volumes and at low prices. Carnegie's genius was to employ the Bessemer process to produce steel on a grand scale. By 1901, the Carnegie Steel Company became J.P. Morgan's U.S. Steel and the versatile ferrous metal established itself as industrialization's most valuable material.

Carnegie first used his steel for railroad tracks. It was a more wear-resistant alternative to traditional iron rails. Carnegie was Ayn Rand's (1905-1982) inspiration for Hank Rearden in her bestselling book *Atlas Shrugged* (1957). The product line quickly expanded as steel mills, aided by rolling mills, offered high-grade steel in a variety of shapes such as sheet, plate, bar, I-beam, and tubing, all available in a wide range of sizes. Engineers and architects discovered that the combination of steel's high strength and its extensive range of standard forms offered endless construction possibilities. Early twentieth-century steel was the critical material for the erection of the world's tallest skyscrapers and longest-spanning bridges. Machine builders relied heavily on steel. Shipbuilders used steel to fabricate hulls and superstructures. Consumer-goods manufacturers employed steel for the bulk of automobile and appliance parts.

In construction, the growing popularity of steel encouraged a surge in concrete production. The Romans used concrete for the construction of large arenas, harbor piers, and monuments, but the technology vanished after the fall of the mighty empire. Europeans reinvented concrete during the 1600s. Centuries later in the United States steel encouraged a twentieth-century concrete boom. Steel skeletal reinforcements complimented concrete forms. Heavy-duty trucks with steel frames and bodies delivered concrete, while construction equipment, which also used large amounts of steel, worked the concrete to build roads, bridges, buildings, and dams.

In their quest to understand steel, chemists introduced the field of metallurgy in the early twentieth century. Their newfound knowledge transitioned the profession from "trial and error" black art to scientific discovery in the latter nineteenth century. Technologies in the black-art phase depended more on knowhow than scientific knowledge. By the 1920s, technical universities taught material science courses at the molecular level. Knowledge and American innovation then led to a surge of modern materials development during the 1930s.

## **Modern 1930s Materials**

After Andrew Carnegie leveraged Bessemer's process to revolutionize the steel industry, steel mills evolved at a rapid rate. The latest steel-making methods, using either open-hearth or electric furnace equipment, produced higher quality steels at faster speeds. By 1930, less than 5% of American steel mills used the Bessemer process. As part of their upgrade, steel mills increased automation reducing the number of man hours required to produce a ton of steel. As a result, the cost of steel fell.

Already a game changer, steel had tremendous potential that modern society needed to the exploit—and that is what they did during the Great Depression. For example, on the finishing side of the sheet-steel business, precision roll machines produced a higher-quality of sheet steel. The new-and-improved sheet metal came in a variety of gage thicknesses, each held to tighter specifications. The higher-grade steel featured a smooth finish, plus characteristics which enhanced the surface for premium painted finishes.

The development of alloy steels became one of the main benefactors of material scientific knowledge. Alloy steels consist of iron and carbon along with one or more metallic elements fused or dissolved into the steel. By tweaking steel's recipe metallurgists increased one or more critical attributes such as strength, toughness, or hardness. Popular alloying elements used in steel include chromium, nickel, molybdenum, manganese, silicon, and tungsten. In 1935, there were already 102 classified alloy steels. By 1939, the fast-moving field of metallurgy increased that number to over 300.

Metal techs further enhanced steel's strength, hardness, and toughness through heat treat processes. Heat treat methods change the molecular structure of steel by heating the metal up near its melting point followed by a rapid cooling process known as quenching. The finished product gave manufacturers the materials to build components that transmitted higher horsepower through compact packages with improved durability.

Stainless steels are another class of alloy steels finding widespread use in the 1930s, and as the name implies, they have corrosion resistance properties, but they are also tough. Stainless steels acquired their unique properties from a large percentage of the alloy, chromium. The shiny steel gained fame in the 1920s as an adornment for structures sporting an art deco flare. During the 1930s industries showcased stainless steel's great utilitarian hygienic value, which became its greatest asset. The metal, often just called stainless, modernized many of the steps and procedures used in the food processing and medical industries.

Aluminum, in part because of lower-cost electricity, expanded its customer base to include dinnerware, appliance parts, automobile components, hardware, packaging, modern furniture and aviation, the latter gaining the most from the light-weight alloy.

# 39

# Tungsten Carbide

Modern consumers demanded products with greater utility and value: more power, increased durability, and of course, lower costs. Those demands encouraged producers to push the limits of advanced designs which brought on new challenges. One of the key technologies tasked with meeting those challenges were cutting machine tools. The next generation of machines were sturdier, more powerful, and produced greater accuracy, but it was their cutting tools that proved to be the weak link. For one, the escalating use of alloy steels in high-performance products demanded high-performance cutting tools. Fortunately, similar alloys served as a superior material for high-performance tooling. By the latter nineteenth century, tooling manufacturers had developed hardened, high-carbon alloy steels, known as high-speed steels (HSS). Their increased hardness reduced cutting tool wear rates when applied to all grades of steel.

Nonetheless, twentieth-century mass production brought more of the same—the need for speed and greater precision—two of industrialization's perpetual objectives. Those demands required high-performance materials and greater precision. The tradeoff for pushing precision technology is its hefty price tag. For example, a common rule of thumb suggests that in many metal-cutting operations halving a tolerance will result in a tenfold cost increase. The use of exotic materials further increased costs. Fortunately, with enough advancing technology, manufacturers can achieve mutually higher precision and lower costs, an arrangement that benefits both producers and consumers. Some of that relief came in the early twentieth century when manufacturers developed cobalt-HSS, an alloy capable of even higher performance. Cobalt's positive qualities included increased hardness compared to standard HSS.

So, the search for higher performance in cutting tools continued. The ideal cutting tool used to sculpture steel must be hard, tough, and heat resistant. Making the selection more difficult is that hardness often degrades with high temperatures, which happens during aggressive, cost-cutting machining cycles. One material not suited to machine steel is diamond, natural or manufactured. Unfortunately, the hardest material known has a negative reaction to heated steel.

High-speed steels provided an incremental step in cutting tool advancements, but it was during the Great Depression that cutting tools made one of their greatest breakthroughs—tungsten carbide. Although unknown to much of today's population, everyone benefits from that unique material.

Carbides are a special mixture of carbon and a heavy metal, in this case tungsten. Tungsten has the highest melting temperature of any metal, 5,000 °F (steel melts at 2500°F). Partly because of its resistance to high temperatures, tungsten became the standard material for incandescent light bulb filaments after 1909. The problem with its application as a cutting tool material is that metal carbides are extremely hard but brittle. A brittle tool will break during high-stress machining.

In 1930, chemists discovered a practical recipe for tungsten and carbide that produced a tool with extreme hardness and heat resistance combined with reasonable toughness—the ideal cutting tool for many industries' toughest applications.

German and American chemists had invented tungsten carbide during the late 1920s with the Germans taking the lead, using it extensively during the 1930s. They had perfected a practical recipe for tungsten and carbide that produced a tool with extreme hardness and heat resistance combined with reasonable toughness—the ideal cutting tool for many of industries' toughest applications. In the United States, credit a metallurgist, Philip M. McKenna, founder of Kennametal, for its domestic introduction. Tungsten carbide, usually referred to as carbide, dramatically increased machining speed and precision. Another advantage to designers and builders of equipment was that carbide tools effectively machined steels and other metals too hard for HSS. Thus, manufacturers switched to stronger, tougher, and harder materials to further enhance their end products' value.

Another tungsten carbide advantage is flexibility. As a sintered metal (a heated powder), tungsten carbide can be formed into complex shapes. After the molding process, modern manufacturing methods give the tooling a smooth finish.

Figure 39.1 illustrates a

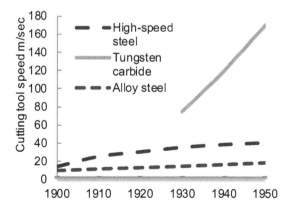

**Figure 39.1** Progression of cutting tool speed, 1900-1950, comparing the ability of alloy steel, high-speed steel, and tungsten carbide to machine plain steel.

metric that reinforces the dramatic advances in metal cutting tool steels between 1900 and 1950. In the example shown, HSS performance tools cut plain steel three times faster during the recovery years compared to their performance by the onset of the Great Depression—tungsten carbide an impressive fourteen times faster.

Albeit expensive for most applications and therefore slow to implement, carbide tools dramatically improved America's ability to produce higher-quality goods at lower costs. Tungsten carbide also served well for dies, tool holders, and petroleum drill tips, along with many other applications. For less demanding applications, toolmakers continued to improve the properties of high-speed steels, giving machinists the flexibility to use the most cost-effective cutting tool for the job.

# 40
# Stick Welding

Iron reigned for much of the nineteenth century as heavy industry's key building material. By the turn of the twentieth century, steel had replaced iron in many applications. One of steel's advantages over iron was its improved weldability. Welding processes provided a practical alternative to typical fastener hardware (screws, nuts, rivets, etc.). Equipment builders also relied on welded steel components as a replacement for iron castings, which required costly molds. Welding equipment matured during the 1930s, becoming one of the valuable tools that helped bridge the gap between older industrial-age technologies and twentieth-century modern marvels.

Welding is a manufacturing process used to bond pieces of metal together by melting. The application of intense heat forms a pool of molten metal along a joint. As the metal cools it coalesces to form a strong bridge (the weld) between adjoining parts. When properly applied, a welded seam becomes stronger than its original members.

Blacksmiths have been welding metals together since antiquity. The laborious process required lots of heat from an open flame and repeated hammer blows to remove any impurities for a strong bond. After World War I, the advancing field of welding covered an extensive range of technologies. Their dynamic names matched their technical sophistication and included oxygen-acetylene (gas), fusion, thermic, hydrogen atomic, and arc. Despite significant progress during the 1920s, the need for complex equipment and highly skilled labor prevented the field from reaching its potential. But one of those technologies, shielded metal arc welding (SMAW), known in the trade as arc or stick welding, was poised to revolutionize American manufacturing on the eve of the Great Depression.

In the early nineteenth century, scientists experimented with a special electrical current designed to melt metals. A century later, the technology evolved into arc welding, another game-changing 1920s technology serving manufacturing.

Arc welding requires a small generator that sends an electrical current through a metal or carbon rod (electrode) placed a fraction of an inch above the weld zone. When the electricity jumps the air gap, it produces an arc capable of heating a metal's surface to temperatures exceeding 6,000°F (steel melts at 2,500°F). The high temperature helps a skilled operator produce a strong, uniform weld at a rapid rate.

While electricity solved the welder's problem of ample heat, it did not address welding's other challenge—contamination. A proper weld needs to be protected from atmospheric gases, particularly oxygen and nitrogen. If a weld absorbs those gases, it becomes weak and prone to failure. Manufacturers experimented with various techniques of protecting the weld, but with only modest success. The big breakthrough came in 1929 when a welding equipment manufacturer company, Lincoln Electric, introduced its mass-produced, flux-coated electrode. The stick, about 5/32 inches in diameter and 14 inches long, had a special coating that when heated emitted an inert gas. The gas formed a protective coating around the weld called slag. Slag absorbs outside contaminates and shields the weld until everything solidifies.

Shielded electrodes complimented the electric welder and came in a variety of materials, each designed for a specific application. Even with an adjustable generator and flux-coated rods, welding still required great skill; however, it was a straightforward, learnable trade.

The army of 1930s artisans equipped with modern arc-welding tools revolutionized manufacturing. They produced consistent-quality welds, yet they maintained high precision with great speed. This novel construction technique transformed a wide range of mechanical structures such as ships, bridges, buildings, containers, oil rigs, and vehicles for rail, road, and off road. In each of those applications, welded steel made the structures stronger and lighter but less expensive.

**Figure 40-1** Steel weldment for machine base top plate shown as transparent

Stick welding produced another ground-breaking manufacturing building block called a *weldment*, a type of steel fabrication that is lighter and stronger than castings or steel subassemblies attached with fasteners. Weldments have the additional advantage of customization compared to standard iron castings. A weldment is an assembly made from an assortment of steel mill components (bars, plates, tubing, I-beams, etc.)—all welded together.

As another technology offering widespread benefits, weldments provided every industry in the country with a valuable manufacturing tool.

Arc welding played a critical role in the massive World War II military industrial machine, particularly in the construction of the prodigious numbers of navy and cargo ships. The US government purchased over 150,000 welding machines for the war effort and then trained a large workforce to operate them. Women made excellent welders, as symbolized by their version of Rosie the Riveter, Wendy the Welder.

Welding technologies helped expand commercial shipbuilding before and after the war leading to the design of safer ships. When divers examined the RMS *Titanic* (launched in 1911) as it lay on the ocean floor, they discovered the iceberg popped many rivets used to fasten the hull's steel plates together, exposing the luxury liner to massive flooding. A 1930s welded hull may have survived the collision.

A "Wendy Welder" at the Richmond Shipyards-1943. *Courtesy U.S. Office of War Information*

Other welding innovations developed during the 1930s made a significant impact on the manufacturing of both private sector and military equipment. Automakers relied on resistive spot welding for fabricating car bodies. Submerged arc welding (SAW), first used in 1935, offered several advantages over conventional arc welding despite the challenges of welding under water. SAW used a continuously fed electrode with a heavy coating of flux. Also, in the mid-1930s, shipbuilders used stud welding for fastening wood decks to steel hulls. By 1941, manufacturers perfected tungsten inert gas (TIG) welding. Where arc welding worked well for thicker-walled steels, TIG welding had a delicate touch, ideal for thin steels, stainless steels, and lighter materials, particularly aluminum.

In a reflection of their own internal manufacturing advancements, Lincoln Electric's history site notes that the cost of their flux-coated welding rods dropped from 16 cents per pound in 1929 to 6 cents per pound in 1942. In contrast, during that same period, Lincoln states that the wage rate for their employees more than doubled. This cost-to-wage example, a nearly 6-fold advantage, offers insight into the wonders of innovative manufacturing usually masked by inflation.

Countless fabrications benefited from the cost reductions and added robustness of modern welding, playing their part to boost the proficiency of a large chunk of the Great Depression economy.

# 41
# Inhaling Freon

As the twentieth century unfolded, scientists, inventors, and manufacturers teamed up to introduce a multitude of revolutionary technologies, which soon became can't-live-without necessities. Near the top of that list must be mechanical chilling machines. Whether cooling a stifling-hot room, keeping food fresh, or making ice cubes to chill a beverage on a sweltering summer day, consumers eagerly embraced air conditioning units and refrigerators.

The science of air conditioning (A/C) and refrigeration emerged after decades of development. This research led to the invention of the refrigeration cycle, a thermodynamic system that uses mechanical energy to make one space cooler by removing heat and dumping it into another via a volatile fluid.

In 1902, Willis Carrier (1876–1950) from Buffalo, New York, invented the first air-conditioning system. Carrier's system not only controlled indoor summer temperatures but also reduced humidity, a benefit for human comfort and for some factory processes. Carrier's first customers were manufacturers whose products were sensitive to heat and humidity, such as cotton textiles and printing.

A decade later, in 1913, Fred Wolf (1879–1954) from Fort Wayne, Indiana, introduced the first home electric refrigerator. Within a few years, two dozen manufacturers competed in the domestic refrigerator market. The two most successful companies were the Kelvinator Company, founded by inventor Nathaniel Wales (1883–1974), and the Frigidaire Company, established by entrepreneur Billy Durant, founder of General Motors.

Both sides of the industry, air conditioning and refrigeration, matured during the 1920s. Carrier first mass produced their patented commercial air conditioning system for industry, stores, theaters, office buildings, museums, and luxury homes. Meanwhile, the number of manufacturers building household refrigerators increased.

By 1930, total US production of refrigerators passed the one-million mark, but at twice the price of a Model T Ford, a refrigerator remained a luxury. A/C systems were also a long way from their market potential. A lack of performance by a single component of the refrigeration system, the refrigerant, hampered the industry.

The refrigeration cycle depends on the powerful thermodynamic principle of latent heat. Latent heat defines the energy required to change a substance's phase while maintaining a constant temperature. For example, the conversions of melting ice to a liquid or boiling water to steam are two examples of latent heat at work.

Earlier twentieth century systems used an absorption system which required heat to increase the pressure of the refrigerant gas. By the 1920s, the modern system of mechanical refrigeration matured—a system consisting of a compressor, an expansion valve, the refrigerant, and lots of plumbing, all tuned to create the ideal thermodynamic process. The refrigerant circulating through the system is the element that moves heat from one location to another, for example, from an icebox into a room or from a room to the outside air.

To be practical, a refrigerant must be able to change from a gas to a liquid (condense) at ambient room temperature for a refrigerator, or warmer outside temperatures for A/C units. The refrigerant must also be able to convert back to a gas (evaporate) with the aid of the compressor. The refrigerant candidates prior to the 1930s which included ammonia ($NH_3$), methyl chloride ($CH_3Cl$), sulfur dioxide ($SO_2$), and methyl formate ($C_2H_4O_2$). Each of these options created hazards if they sprang a leak, including flammability, foul smells, or toxic gas. Ammonia also had the disadvantage of being caustic.

For safety reasons, refrigerators needed to be placed in a well-ventilated storage room. Cautious homeowners kept their refrigerators outside. The same cautions applied to air conditioning systems.

In 1928, a trio of corporations, DuPont, General Motors, and Frigidaire, combined their resources toward the common goal of producing a safe yet practical refrigerant. Two years later, in 1930, the consortium formed the Kinetic Chemicals Company to launch their new product, a chlorofluorocarbon (CFC) marketed as Freon. Its most popular derivative was Freon 12, also known as R12. Freon, as a liquid or a gas, is nontoxic, odorless, nonflammable, and noncorrosive. Only ammonia has more ideal specifications than Freon, finding applications in large refrigeration systems, but that advantage is slim.

One of Freon's primary inventors, Thomas Midgley Jr. (1889–1944), once made a dramatic demonstration of the gentle nature of the new refrigerant. In front of an attentive audience, Midgley filled his lungs with Freon and then exhaled the gas over a candle flame. The flame went out, while the scientist suffered no ill effects from inhaling the Freon.

The implementation of the electrical grid in the early twentieth century gave a huge boost to the commercial potential for A/C and refrigeration. Electricity supplied the horsepower for the motor-driven compressor, and the means to build a simple yet reliable automatic control.

Air conditioning giant, the Carrier Corporation, switched its primary refrigerant to Freon in 1932, just in time for its burgeoning commercial customer base—movie theaters and office buildings. As a refuge from the troubles of the Great Depression, moviegoers flocked to theaters. Air conditioning on a hot day added to the thrill of enjoying a talkie during the Golden Age of Hollywood. An increased focus on general bookkeeping methods and other administration functions, from both the private and government sectors, drove the construction of air-conditioned office buildings.

Freon complimented refrigerator design improvements and price cuts to propel vigorous sales. From 1930 through 1935, during the deepest depths of the economic panic, appliance makers sold eight million refrigerators, most operating with a Freon refrigerant.

By the 1980s, scientists concluded that CFCs from aerosol cans or leaking refrigerators and A/C units were destroying the Earth's vital ozone layer. In response, governments from around the world began to restrict the use of CFCs through numerous international agreements. But in 1932, Freon was just one in a long list of miracle materials introduced during the Great Depression—another technology that pushed the struggling economy along and provided yet more incentives to purchase during the eventual buying frenzy that followed the war.

# 42

# The Plastics Revolution

Another set of remarkable inventions introduced during the Great Depression came from a family of manmade synthetic polymers developed by America's largest chemical companies. Their inventors assigned them scientific trade names such as polytetrafluoroethylene, polyvinylidene, and polyamides. Collectively, we know these materials as plastics. We are more familiar with some of their generic names and registered trademarks, including neoprene (DuPont 1931), fiberglass (Owens-Corning Fiberglass Company, 1935), Plexiglass (Rohm and Haas, 1936), nylon (DuPont, 1937), silicone (General Electric, 1940), Saran (Dow Chemical Company, 1943), and Teflon (Kinetic Corporation, 1945). These popular products, and many more, came directly or indirectly from breakthroughs completed during the 1930s. This decade, while not known as an innovative period, produced an impressive percentage of modern polymers still in use today.

By definition, a plastic is a material shaped by heat and pressure and then hardened. Polymers are plastics consisting of repeating strings of large molecules. Their unique molecular chains give the material its light weight and a wide range of attributes.

Mankind's use of plastics dates back thousands of years and includes natural compounds such as shellac and eggs. The earliest manmade plastics came from England in the mid-nineteenth century, most famously as a replacement for ivory billiard balls.

John Wesley Hyatt (1837–1920) invented the first American plastic, celluloid. Celluloid found use in early motion picture films. In 1907, John Baekeland (1863–1944), a Belgian immigrant living in Yonkers, New York, developed another successful synthetic plastic, Bakelite. Bakelite resembled wood, but did not shrink, split, or warp. Although popular in the burgeoning electronics industry, it was used in many other items ranging from buttons to gears.

By the turn of the century, chemistry had developed into a glamorous, high-technology field of study offered by universities. Later, the major chemical companies' research labs hired chemistry graduates to experiment on a new family of polymers based on coal or petroleum. The research paid off leading to an explosion of innovation, including many important discoveries made by accident. Plastic's unique characteristics, which combined versatility and low manufacturing costs, strengthened America's capability to produce valuable goods.

Plastic's adaptability made it practical as a substitute for numerous manufactured and natural materials in common use including metals, wood, textiles, glass, ceramics, leather, ivory, and rubber. All plastics have the often-desirable characteristics of light weight, high dielectric strength (electrical insulation), and corrosion resistance. Chemists customized polymers to achieve a wide array of properties such as a specific flexibility or hardness. They made plastics that were transparent, translucent, or opaque, and in a variety of colors and patterns. Researchers developed recipes to make polymers with surface textures that could be slippery or sticky.

Plastic's greatest advantage came from its ease of manufacturing. Whether pressed, extruded, machined, or molded, the polymer industry's machinery pumped out plastic parts faster than their traditional alternatives. Injection molding, a process integrated into a new generation of machines introduced in the mid-1930s and modernized in 1947, offered some of the largest gains from polymers. Compared to high-production metal processes such as stamping and die casting, injection molding machines ran at much higher speeds, and they could produce significantly more complex parts with minimal waste and a smoother finish.

DuPont's nylon was the first synthetic polymer of the 1930s class to become a huge commercial success. DuPont chemists developed nylon as a replacement for a wide range of materials including silk, a valuable textile then in short supply because of escalating tensions with one of its chief exporters, Japan. Released in 1938, nylon was first used for combs and toothbrush bristles. Nylon's most famous application came as an alternative to lady's silk stockings. DuPont launched their miracle fabric at the 1939 World's Fair held in New York City. Consumers could not buy enough nylon stockings. As noted, DuPont sold 64 million pairs in its first year of production. During World War II, the manufacturing of nylon shifted to military applications, most notably, parachutes and tents. After the war, the love affair with nylons returned, as women by the millions waited in long lines to purchase synthetic sheer stockings.

Aesthetics is another one of plastic's valuable attributes. In the late 1940s, jukebox manufacturers such as Wurlitzer, Seeburg, and Filben Maestro produced striking, modern designs highlighted by liberal amounts of plastics in conjunction with brilliant, multi-colored backlighting. Today, these jukeboxes are highly collectable.

Record companies made 78 RPM records out of vinyl starting in 1931 as an alternative to traditional shellac records. In 1939, RCA introduced 7-inch 45 rpm records with narrow microgrooves. Vinyl offered the advantages of lighter weight and improved sound quality, and they were less likely to break if dropped. On the minus side, nylon records attracted dust through static electricity and the heavy stylus arms of the day caused premature wear. After the war, to accommodate the special characteristics of softer vinyl records, manufacturers developed record players with lightweight arms using sapphire styluses.

Plastics found another niche in the toy industry. Manufacturers favored its ease-of-manufacturing, which translated to lower costs. Consumers bought plastic toys for their affordability, but also valued those soft edges for a child's safety.

As a lightweight and low-cost material, plastics were perfect for packaging; by design a throwaway product. Initial uses included bottles and cosmetic cases. That was just the beginning. Due to their enormous potential, the customer base for the plastic packaging industry grew rapidly. The fact that plastics were not biodegradable was not an issue at the time.

The electronics industry applied plastics for both hardware and cabinets. Radios were already a mass-produced item throughout the Great Depression, perfect for high-production plastic processes. After the war, the popularity of televisions soared, providing another large market for polymers.

Automakers first looked to plastic for dashboard knobs, later employing them for much of a car's interior. Synthetic rubber tires improved durability in both cars and trucks.

Romex electrical cable, invented in the 1920s, reduced the time to wire anything in the construction industry, whether residential, commercial, industrial, or military. The original Romex cable used tarred cloth and paper to insulate the copper wires. During the mid-1930s, the company switched to PVC, which lowered costs and improved durability.

Plastics, while featuring numerous positive qualities, have their share of weaknesses. Some early plastics had low strength, were susceptible to heat, and others emitted a strange odor. As a result, when designers employed polymers incorrectly, or in a less than ideal situation, the unfortunate results led to the phrase *cheap plastic*. Historians tend to focus unfairly on those failures while ignoring plastic's greater contribution. For example, social commentator and Pulitzer Prize-winning author Norman Mailer (1923–2007) had this to say about plastics in the *Harvard Review*:

> I sometimes think there is a malign force loose in the universe

that is the social equivalent of cancer, and it's plastic. It infiltrates everything. It gets into every single pore of productive life.

The plastic revolution did not always impress academics either. Donald Cardwel(1919-1998)1, a professor specializing in the history of technology, wrote in his book Wheels, Clocks, and Rockets: A History of Technology (2001) wrote:

> Although plastics represent a relatively new technology with enormously wide and numerous applications, they can hardly be classed as a strategic technology.

In fact, the bulk of texts reflecting on the history of the period simply omit any discussion of the explosive development of polymers.

While everybody has their frustrating moments when poorly designed plastic parts fail, those instances represent a small percentage of application. The excessive criticisms by pundits for a material beneficial for the majority of consumers smells of elitism. The new group of plastics developed in the 1930s complimented American mass-manufacturing and allowed people of the lower-income classes to purchase more goods than they could have in the pre-plastic days.

The burst of innovation created families of new-age materials that appear to be ahead of their time. This does not fit conventional wisdom's view of Great Depression technology. Plastics are completely engineered. First tuned in the laboratory for a specific use, plastics then moved to the manufacturing floor where they fulfilled a wide breadth of applications. The fact is, polymers and other manmade plastics developed during the 1930s made a significant impact on the economy. A few designs became instant successes, particularly nylons. They arrived just in time to assist in the war effort where they merged in with other hi-tech materials. The recovery needed the cycle of consumption and production to gain momentum—incredible plastics, with their low-cost and unique properties, contributed to that stimulus. During the 1950s, plastics turned into an enormous industry helping to assure the economy's long-term success.

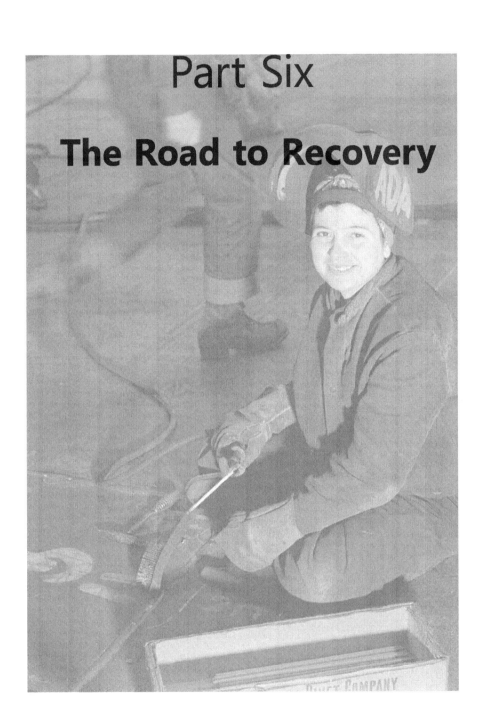

# Part Six

# The Road to Recovery

## 43

# World War Two Interrupts the Great Depression

By 1940, the US economy was prime for a recovery. The nation's industrial transformation continued to gain momentum. Enthusiasm for the progressive revolution was dwindling. In the alternative exercise (no World War II), the recovery sticks a few years earlier—between 1942 and 1944. The New Deal's reputation gets a boost and sole credit for the recovery—a congratulatory "job well done" from academia with much of that praise going to FDR. Critical programs remain in place, some to keep *devious* capitalism in check, others to prevent another Great Depression, and a few more to serve as safety nets for society's most downtrodden citizens. The big unknown, compared to the actual post-war economy—would a peacetime recovery have been stronger, weaker, or similar? Either way, the New Deal gets the credit for ending the Great Depression. But history took a detour and America went to war.

As is the case with the 1929 crash or the ensuing long economic slide, there is no ironclad explanation for the latter-1940s recovery. As noted, if Germany and Japan had behaved themselves, troubleshooting the Great Depression's recovery would have been less speculative. Nevertheless, there are technological and political trends that provide historian sleuths with sufficient clues to form a reasonable consensus as to how the Great Depression ended.

A useful clarification for the war's role in the recovery is to ignore the claim that WW II ended the Great Depression and replace it with the more pragmatic claim—WW II interrupted the Great Depression. Of course, mainstream pundits' misleading conclusions claiming that the war kicked off the recovery have had enormous success. Their argument relies on two well-known wartime statistics. The unemployment rate plummeted to record lows. The GDP, economists' dubious growth metric, skyrocketed. Plus, the timeline fits. The economy rebounded within a reasonable timeframe after the war ended. Nevertheless, the simplicity of the tenet that "World War II ended the Great Depression" understates the complexity of the era.

The low unemployment rate and high GDP have solid assignable causes which are not recovery related and should be dismissed as part of any positive, systemic change. GIs lowered the unemployment rate. An influx of military spending increased the number of jobs and propelled the GDP. Another fallacy understates the many high-tech tools and technologies used during the war that came from the 1930s.

There is an interesting contrast between real prosperity and economic statistics for the 1930s compared to the early 1940s. During the 12 years leading up to the war the American economy excelled despite the record high unemployment rate. Wartime had the opposite effect. Throughout the war years, the economy suffered despite the record low unemployment rate. In both cases, the extraordinary numbers did not last long. Once the country returned to a civilian economy, the unemployment rate shot up and the GDP plunged. Other complex issues add suspicion to the "war ended the Great Depression" viewpoint.

The Allies' victory did not bring about an immediate economic expansion. Instead, the war, followed by the peacetime conversion, put on hold the resurgence of a modernized private sector already bursting at the seams to spend. The New Deal's social band-aid and disaster relief stimulus packages continued to wind down while its burden on the economy after the war was far from over. FDR, the progressive movement's charismatic leader and motivational force had been replaced by the less-popular Harry S. Truman. Finally, the efforts involved in transitioning industries from wartime to peacetime production continued to inflict economic turmoil deep into the 1940s.

The war's effect on the Great Depression aside, America's industrial defense efforts turned into a momentous historic event. The effort by hard-working American men and women to build a formidable military-industrial complex virtually from scratch impresses anyone who studies the period. For the war effort, Americans unleashed a ton of innovation to further separate the world's number-one industrial superpower from the rest. This rapid launch provided a timely response against America's enemies with a wide range of sophisticated military equipment, manufactured in high volume. Many of those technologies boosted industrial productivity after the war.

✧✧✧

American defense expenditures had been anemic since the end of WWI—the war to end all wars. By the late 1930s, Nazi aggression had persuaded American politicians and military leaders to reexamine the nation's military-isolationist agenda. Their immediate concern was that America's close ally Great Britain would be unable to fend off an impending German invasion. In response, FDR ordered an increase in the production of armaments.

That was a start, but the free world's adversary was no lightweight. Adolf Hitler (1889–1945) had invested heavily in hi-tech armaments. His rapid conquest of Poland, followed by Norway, Denmark, and France, along with air and sea attacks against England, proved to not only validate the new Germany's military might but also its hostile nature.

Roosevelt answered back with a more aggressive plan. US military commanders added to Roosevelt's alarm. They pleaded with the President to increase the government's military investments—significantly. FDR agreed. To develop a catch-up strategy, President Roosevelt, in May 1940, assembled a group of industrialists led by the renowned automotive executive Bill Knudsen (1879–1948). Their starting point had few specifics except one strong message—increase America's military capabilities.

The urgency of pre-war armament manufacturing escalated again after March 11, 1941, when Congress passed An Act to Promote the Defense of the United States, better known as the Lend-Lease Policy. The law created *loophole* legislation bypassing American isolationist laws adopted after World War I that forbade the subsidizing of foreign wars. The program, which was palatable to those wanting the US to remain neutral, *loaned* highly subsidized military goods to enemies of the Nazis with the bulk going to Great Britain and the Soviet Union.

After Pearl Harbor and Germany's declaration of war against the US in December 1941, Roosevelt pushed America's military manufacturing capacity to its limits. FDR commanded his team to direct much of the USA industrial machine away from commercial production and toward the construction of military hardware. Hitler's advisors considered the grand plan American propaganda and told their *Führer* not to worry—the United States was more fluff than substance. Japan's Admiral Isoroku Yamamoto's (1884–1943) famous assertion, "I fear all we have done is to awaken a sleeping giant..." turned out to be one of history's greatest understatements.

Once established, factories undershot FDR's earlier goals, but not by much. The speed at which the USA mobilized high volumes of sophisticated military gear, converting anything technology-related into war production, challenges the wildest imaginations. Between 1940 and 1945, American industries built over 100,000 tanks, 250,000 pieces of artillery, 2.3 million vehicles, 2.7 million machine guns, and outfitted 16 million soldiers, aviators, sailors, and marines.

The expensive defense equipment operating in the skies and on the seas was all the more impressive. From 1941 through 1945, American factories built 325,000 airplanes, 97,000 in 1944 alone. The prodigious number of planes included 15,000 P-51 Mustangs, and 15,600 P-47 Thunderbolts. The versatile escort-fighters served well in both dogfights and bombing raids. The US constructed 18,000 of the larger B-24 heavy bombers, half of them made at the Ford Motor Company's Willow Run, Michigan, special-purpose factory. They built the largest building in the world so that at peak production in 1944, Ford could produce a B-24 every 55 minutes. Those, along with other bombers such as the B-17 and B-25, inflicted enormous damage on enemy troops and their infrastructures.

An even larger bomber, the B-29 Superfortress, challenged aircraft builders on both its size and technology. A fully loaded B-29 weighed 60 tons yet had a range of 3,250 miles. The Superfortress was the first production aircraft with a pressurized cabin and analog computers, which operated remote-controlled machine gun turrets. The computers calculated trajectories, accounting for variations in airspeed, temperature, and humidity. The B-29, as with other bombers, had gun canopies, or blisters, made from revolutionary Plexiglas. The US assembled 3,000 B-29s. All served in the Pacific theater, including the bombing of Tokyo and the dropping of atomic bombs over Hiroshima and Nagasaki.

The build-up of the Navy required a greater tonnage of material and a larger labor force. At the outbreak of the war, the United States had a small, aging fleet of ships, yet it had to fight enemies protected from American forces by two vast oceans. Within a couple of years, America had created by far the most powerful Navy in the world. The wartime production comprised 35,000 landing craft and 890 larger ships. The latter group included 245 submarines, 349 destroyers, 48 cruisers, 8 battleships, and 163 aircraft carriers. Equipped for battle, carriers weighed 35,000 tons, accommodated 100 aircraft, and required a crew of 2,600.

The US needed a significant increase in cargo capacity to transport those armaments to overseas destinations. To accomplish this goal, American industries constructed shipyards to build bare-bones cargo vessels designed for mass production called Liberty ships. Construction magnate Henry J. Kaiser (1882-1967) led this wartime initiative. Equipped with light armament, a Liberty ship could haul 10,000 tons of cargo. In a "build them faster than the Germans can sink them" strategy, Kaiser and his teams constructed 2710 Liberty ships. Toward the end of the war, the merchant force built 531 of larger and faster freighters—the Victory ship.

Today, political conservatives place the blame for the extended length of the Great Depression squarely on the shoulders of the three-plus-term president, Franklin D. Roosevelt. In contrast, FDR's bold strategy and strong leadership during World War II brought a round of applause and redemption from many of his Republican critics. The President's aggressive approach ended the war against formidable foes in fewer than four years. America's military strength led to the war's short duration and a decisive victory. The war's unconditional surrenders turned our once mighty foes into powerful allies and enabled a relatively speedy return to normalcy. A longer war would have cost more lives and buried the United States deeper in debt, delaying the recovery.

Modern scholarship considers America's industrial war machine not only a successful military strategy but also, confirms that stimulus (printed money) rejuvenated factory production, and in time, the economy. Therefore, the two must be related. As often happens with Great Depression scholarship, beyond conventional wisdom's jubilation lies another side to the story.

America learned to build the World War II military machine from 1930s manufacturing advancements. Factory modernization, machinery technology, steel and aluminum production, other material sciences, electronics, aircraft expertise, and propulsion horsepower all grew stronger in the years preceding the war. Without the burden of the New Deal, that technology could have driven a robust recovery.

The term *recovery* covers a considerable range of economic growth. To its advantage, the American economy had enormous industrial and commercial capacity, plus the US Constitution took some hits, but overall survived a decade and a half of radical political challenges. To avoid a recessionary relapse, the American people would have to rediscover their conservative roots.

# 44

# The Catalysts Driving the Recovery

The Great Depression economy operated by an age-old tenet—citizens worked to purchase goods and services. During the 1930s that system functioned to a point. High unemployment and sub-standard productivity lingered because superimposed over traditional commerce were the efforts of progressives out to crush the American spirit. Their commitment to excessive taxation, overbearing regulations, socialized labor, and twisted anti-monopoly laws built a wall that limited American exceptionalism. For a solid recovery, the nation's economy needed a break. There was no need for pols to revert to the days of the Coolidge administration, but left-leaning ideologues needed to back off. The future of American prosperity hung in the balance.

The challenges of war preparation created some of those political-economic positives—a taste of laissez-faire. Compared to his previous 7 years in office, the Commander-in-Chief had to learn how to play nice with industry. As a result, Roosevelt employed a distinct set of rules for refereeing the private sector during the war as opposed to his treatment of them throughout the 1930s. First, following America's declaration of war, the President arranged for organized labor leaders to outlaw labor disputes. There were many unauthorized, smaller wildcat strikes, but union leaders, in step with the President's mission, prohibited *unpatriotic behavior* and kept their members working.

FDR's second war concession, and another deviation from the Democrat's progressive agenda, was to order the Department of Justice to back off their constant barrage of frivolous criminal charges against corporate excellence. If the federal government had shifted to a pro-industry position in the 1930s, would the Great Depression have ended sooner?

War placed much of the New Deal in a state of flux and commerce on hold. Parts of it were ending while a potential fresh round of initiatives loomed on the horizon. Wartime demands reduced or halted the production of many civilian goods. For the time, the war and not the economy consumed the full attention of the American people. Then, suddenly it was over. Time to rebuild.

In the nation's 1937 upstart attempt, the economy began its third rejuvenation process (the first was the brief 1930 recovery and the second was the 1933 FDR boost). It had been a long, seven-plus years since the start of the Great Depression. To recap. That year the unemployment rate dropped to near 14%, its lowest level since 1931, and consumer optimism inspired a healthy jump in private-sector spending, which approached 1929 levels. On Wall Street, the Dow peaked at 190 on August 16, 1937, the highest average stock price since February 1931. Then, during the summer of 1937 the economy tumbled. Instead of a true recovery, the resurgence turned out to be another economic bubble that burst. The crash also coincided with a new round of regulations, spending, and taxation that curbed economic growth and drove the unemployment rate up to 18% breaking the enthusiasm of consumers and investors. The Dow fell 82 points (43%) in under nine months. By 1940, a partial recovery had just two years of traction, but with a global war looming, politicians would have to place economic growth initiatives on hold.

In 1945, the American economy would make its final Great Depression attempt at an economic revival. The private sector had built up enormous amounts of economic potential, well beyond its capability during the decade's previous bid at a recovery.

The post-war economy spawned a political battle zone—socialism versus the free market. On the destructive side, progressives were still in charge and large components of the New Deal remained intact. But there were powerful catalysts—an eclectic assortment of political, military, and cultural events—assisting a post-war recovery. The transformation encouraged a modernized private sector way overdue for a dramatic expansion.

## Free Trade

The 1930 Smoot-Hawley Tariff Act damaged the economy. Therefore, it makes sense that a reversal of a punitive trade policy could make for a powerful recovery tool, and it did.

The US government commanded high tariffs during the 1920s, extremely high tariffs during the first half of the 1930s, yet low tariffs by the 1940s. An enlightened and historic trade policy signified one of the rare constructive political policies to emerge during the pre-war Roosevelt administration. A profound economic bonus, America's new liberal trade policy contained enough power to offset much of the damage triggered by lingering New Deal programs.

The Smoot-Hawley era marked a low in modern US trade policy. The first major step toward reversing those high punitive tariffs came in 1934 during FDR's first term. That year, on June 12, President Roosevelt signed into law the Reciprocal Trade Agreement Act (RTAA). Congress designed the RTAA to reduce the high tariffs enacted through Smoot-Hawley in the short and long term. One tool granted in the RTAA gave the President the authority to negotiate bi-lateral trade agreements, a potent power normally assigned to

Congress. For other non-treaty tariffs, the trade act made it easier for Congress to lower tariffs with a simple majority vote instead of the prior standard—a two-thirds majority. From that point, American trade policy followed an unprecedented path to open trade.

Cordell Hull (1871-1955), FDR's long-serving Secretary of State, was the driving force behind America's conversion to a free-trading nation. A lifetime Democrat, Hull served in the Spanish-American War, in the US Congress and Senate, and received the Nobel Peace Prize in 1945. Not one of FDR's favorites, Hull remained popular among the American people. To his credit, the President opted for pragmatism and kept him onboard. Today, his popularity has waned for his refusal to allow freedom ships carrying European Jews fleeing the Nazis entry into the United States. Eventually, the First Lady and the Undersecretary of State overruled Hull's policy but not before previous directives forced the refugees back to Europe where the Nazis captured them. The bulk of those prisoners died in Hitler's concentration camps.

Despite his liberal pedigree and the Jewish refugee tragedy, Cordell Hull stands out as a freedom warrior amongst the crowd of his fellow progressives. With few allies, Hull soloed his way through his free-trade crusade. His 11 years of persistence paid off. By 1939, the United States had 20 free-trade agreements with foreign nations. Hull lacked partners, but current events, such as the lend-lease program, America's prominent role in World War II (which put any Smoot-Hawley animosity toward the US aside), and later the new United Nations promoted a global atmosphere of cooperation and free trade.

Thanks to Cordell Hull, tariff rates declined throughout the Great Depression encouraging free-trade momentum after the war. In 1947, the United States signed the General Agreement on Trade and Tariffs (GATT), a pact that included 28 nations. Other GATT treaties followed, helping to ensure an extended and robust recovery.

Figure 44.1 tracks America's imports and exports from 1921 through 1950. The third line shows the duties collected as a percentage of all imports. Imports and exports doubled during the latter 1940s compared to the late 1920s, which corresponds to the reduction in import duties.

One of the most interesting and beneficial long-term political changes to come out of the Great Depression—a rare reversal in ideology between the two political parties was trade policy. The Republicans, from Lincoln to Hoover, advocated higher tariffs to support their political base—manufacturing. During that same period, Democrats opposed higher tariffs to benefit their largest constituent—farmers. During the 1930s, Democrats found a *new best friend*, the large and powerful organized labor. Protecting union jobs from foreign competition became one of labor's top platforms. This made trade protectionism a goal for the Democrats so they lobbied for higher tariffs. Meanwhile, the Republicans, partly due to the devastation caused by Smoot-Hawley, supported lower

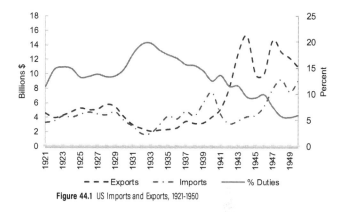

**Figure 44.1** US Imports and Exports, 1921-1950

tariffs. Gradually, the center of gravity of the protectionist trade movement shifted along political agendas. By the mid-1940s, trade ideology among the political parties completed the flip-flop. Fortunately, Democrats were not so much the protectionist militants as the Republicans were prior to the 1940s. This led to a more open trade policy for the United States which was helpful for the recovery and beyond.

Lower tariffs produced a classic laissez-faire chain reaction, benefitting the health of America and its allies' economies. Reduced duties on imports encouraged America's trading partners to reciprocate in kind—the opposite effect triggered by Smoot-Hawley. Also, America's evolving liberal trade policy helped new technologies spread faster across dozens of nations. The result? The economies of our trading partners strengthened, creating an environment through imports that enriched America's economy.

Despite a modest faction of trade protectionists lingering in Congress, by the early 1950s, America's trade renaissance stood as the most open in the nation's history and possibly one of the most tolerant in world history. The USA's newfound enlightened trade policy would have made Adam Smith proud.

## A Slight Move to the Right

Conservatism dominated American politics when the economy collapsed in 1929. Frustrated by the depth of the panic, the political leanings of the nation swung to the political left. When the economic downturn lingered, discouraged voters began to work their way back to the right. This shift affected all three federal branches. Sometimes these moves appeared subtle, as they packed just enough clout to give the private sector the break it needed to propel a dramatic economic recovery.

A rush of voter optimism highlighted Roosevelt's first presidential term, but over time cracks developed between FDR and his constituents. The continued high unemployment rate, which lasted through the 1930s, suggests that the New Deal had not worked as promised. Although FDR won three more elections in 1936, 1940, and 1944, all with a large electoral advantage, his popular-vote margin over his Republican opponents decreased. Roosevelt won 62.4% of the popular vote in 1936, but dropped to 55.0 in 1940, and down to 53.8 % in 1944, relatively low numbers for one of the highest-rated presidents of all time.

FDR envisioned a New Deal sequel—the Super New Deal, but he died on April 12, 1945, just one month into his fourth term. Vice President Harry S. Truman (1884-1972) took over as Chief Executive. Could the new president move FDR's agenda forward?

Bolstered by America's military successes in World War II, Truman's approval rating peaked at an impressive 87% in June 1945. From then on, his popularity plummeted. The President's approval rating was down to 32% by late summer of the following year.

Several factors caused the degradation of the public's admiration for the 33rd president. One reason might be an impatient public with instant gratification needs for a speedy recovery after the war, which was beyond Truman's control. There were also more legitimate issues. Major labor strikes disrupted commerce, and the President took much of the blame. High taxes to pay off the war debt added to America's discontent with the administration. Truman's popularity rebounded in time for a presidential victory in 1948, but nosedived again because of America's disapproval of our involvement in the burgeoning Korean conflict. Truman's approval rating bottomed out at 22%, a record low that neither presidents Richard Nixon (1913–1994), nor George W. Bush (B. 1946) would eclipse during their darkest times.

On September 6, 1945, Truman submitted his 21-point plan to expand the New Deal to Congress. The president named it the Fair Deal. The proposal contained large amounts of price controls, government benefits, and income redistribution. For the Democrats that would have been their logical next step. Paul Samuelson echoed that sentiment. Considered by many to be the Father of Modern Economics, Samuelson led a crusade beginning in 1943 for the federal government to sponsor a massive post-World War II stimulus package. He and other leading economists feared that the shift from military to civilian production, along with 10 million returning service men and women, would crush any prospects for an economic recovery.

The Democrats did not pass anything close to the complete Fair Deal or Samuelson's stimulus agenda. Their lack of action left the recovery some well-needed breathing room. Those blockages allowed for a two-decade delay until in the mid-1960s when former New Dealer, Lyndon B. Johnson (1908–1973), launched his Great Society legislation. That was enough time for the private sector to build a substantial amount of industrial capital.

The Democrats held on to the presidency, whereas a notable political transformation occurred in both houses of Congress. The Graphs in Figure 44.2 depicts these changes. The mirrored shift features some interesting timeline symmetry. In 1929 Hoover joined a Republican majority in Congress. Over the next 10 years both houses moved in favor of the Democrats. At the height of their popularity, during the 1937-1939 term, the Democrats held nearly 80% of the seats in both the House of Representatives and the Senate—a historic majority. After a gradual ten-year reversal, Republicans in the Senate and the House regained their majority. Albeit slight, it was enough and came at a critical moment when America badly needed an influential conservative voice.

✿✿✿

In America's checks-and-balances government, the Supreme Court serves as the last line of defense against over-zealous politicians and their occasional

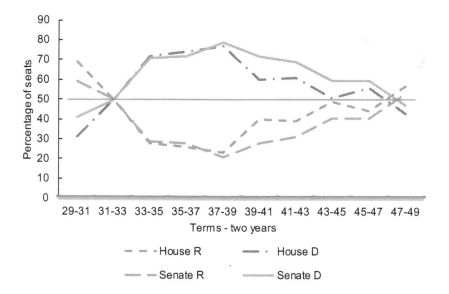

**Figure 44.2** Control of the US Senate and House of Representatives, 1929-1949.

disregard for the United States Constitution. For the Great Depression, the court's decisions on constitutional matters seesawed between a conservative and a liberal bias. While far from being strict constitutionalists, the court did loosely back republican ideals, which may have prevented the Great Depression from taking the nation down a path of even greater ruin.

During the New Deal's early days, the Supreme Court consisted of four conservatives (the Four Horseman), three liberals (the Three Musketeers), and two swing-vote judges. Despite the Four Horseman, that version of the high court backed some of Roosevelt's early New Deal initiatives. When a piece of legislation's constitutionality came into question, the top court sometimes played the "we must save the nation from economic despair to preserve our democracy" card. For Roosevelt and his New Dealers, they appeared to have an ally in the Supreme Court allowing them to get away with quite a lot. Until 1935.

Chapter 30 chronicles the abuses of the New Deal ace, the National Recovery Administration (NRA). This chapter examines the repercussions of shutting the overbearing agency, followed by a chain of events that shifted the Supreme Court's political viewpoints slightly to the right.

According to Amity Shlaes, author of *The Forgotten Man*, the American people owe an enormous debt of gratitude to four butchers from Brooklyn, New York—the brothers Aaron, Joseph, Martin, and Alex Schechter. The Schechter's butcher shop purchased, processed, and sold chickens locally. In 1934, bureaucrats at the National Recovery Administration found time to oppose the way the small business operated, eventually targeting them for federal violations. Auditors from the NRA accused the brothers' kosher chicken packing business of violating the agency's Code of Fair Competition in the Live Poultry Industry. Later that year, at their trial, a federal New York court convicted the brothers for sixty NRA code violations including failure to follow minimum wage and maximum hours rules. The federal court also found the butchers guilty of straight killing. Straight killing allowed customers to select individual live chickens from a coop. The NRA code, created by a federal agency, specified that customers of a local butcher could only purchase chickens by the whole or half of a coop.

The continued harassment forced the brothers out of business, and the felony conviction brought the threat of heavy fines and jail time. Persistent claims by the plaintiffs that the NRA's federal agents used a heavy hand against a small business caught the attention of the Supreme Court. In May 1935, the Court justices heard arguments in the case officially known as *A.L.A. Schechter Poultry Corp. Vs United States*. Federal prosecutors assumed the Hungarian immigrants, with their lack of a formal education and struggles with the English language, would make for an easy victory. Instead, the brothers made the federal bureaucrats look like fools. The Schechters demonstrated that the Jewish dietary law of *shochtim* guaranteed healthy chickens, while the

Blue Eagle that tried to regulate them knew little about the industry they sought to control.

In a huge blow to Roosevelt's New Deal, the judges handed down an unprecedented unanimous decision in favor of the Schechter brothers. The Supreme Court ruled that the NRA abused the Constitution's Commerce Clause, which granted the federal government the right to control commerce between the states. The ruling led to the release of more than 60 individuals under investigation by the NRA for code violations. Nor did the judges simply scold the NRA—instead they declared the entire agency unconstitutional, along with all other agencies created by Congress under the 1933 National Industrial Recovery Act (NIRA), and ordered them all shut down.

The 9-0 decision by the high court sent out a strong message, adding to its historical significance. Each of the Supreme Court judges backed the importance of the United States Constitution over what Congress and the President believed was right for the people. Progressive Supreme Court judge Louis Brandeis (1856–1941) expressed those strong feelings to White House lawyers Tommy Corcoran (1900–1981) and Ben Cohen (1894–1983) when he said, "This is the end of this business of centralization, and I want you to go back and tell the president that we're not going to let this government centralize everything."

The nine justices did not stop with the NRA. They were on a roll. In another 1935 landmark decision, the Supreme Court declared the Agricultural Adjustment Act (AAA), the farm equivalent of the NRA, unconstitutional, and again the judges overruled a key Roosevelt New Deal initiative with a 9-0 decision. In that same year, the New Deal lost a few more decisions, although those were along political lines and typically fell to a narrow 5-4 majority.

Scholars normally evaluate a president's performance by their accomplishments and occasionally their abstentions. However, there is a third, untapped category—what the president tried to do that was foiled by one of the two other remaining branches of the federal government. The ordered shutdown of the NRA should count as a big black mark against Roosevelt's legacy, as the Supreme Court prevented an even larger long-term economic disaster. Until Congress created new legislation to cancel the intrusive regulations, Americans would have suffered under a command economy hovering somewhere between 1960s European socialism and Soviet communism.

Unfortunately, the court's spirited backing of the Constitution did not last long. Feeling the wrath of FDR's anger over their recent rulings, the Supreme Court allowed a string of New Deal initiatives to stand. In the first epic case regarding the easing of constitutional principles, the court backed Washington

State's controversial minimum wage law. The justices then upheld the legality of progressive cornerstones, the Social Security Act and the Labor Reform Act.

President Roosevelt's latter 1930s court victories can be attributed to the jurisprudential shift of swing-judge, Owen Roberts (1875–1955). The press, in describing the Supreme Court's sudden change of heart, modified a classic aphorism to, "the switch in time that saved nine." However, the real political drama had yet to begin.

Despite the Supreme Court's return to backing the New Deal agenda, the earlier 1935 betrayals by the court still angered Roosevelt. He particularly felt animosity toward the Four Horsemen, or as FDR called them, "The Four Old Men." In response, the Roosevelt team crafted a bill designed to mitigate the obstacle, titled the Judicial Procedures Reform Bill of 1937. If it became law, the Judicial Reform Bill would allow the President to add one Supreme Court Justice for each member who had at least 10 years of experience and was over 70-years old. The proposed law limited the number of newly appointed judges to six. If FDR had his way, the Supreme Court theoretically could grow from nine to fifteen judges. The legislation also allowed the sitting president to add another 259 federal judges to the lower-court system. The scheme became known as the court-packing plan and developed into a major policy blunder by the normally politically astute president.

Amid the political backlash, FDR presented his case to the American citizens on March 9, 1937—the first fireside chat of his second term. The President dedicated the entire forty-three-minute radio broadcast to the defense of his impending legislation and denied that the plan was *packing the court*. Roosevelt systematically laid out a defense that the administration had saved the economy but often only by a slim 5-4 majority. FDR complained that, "The change of one vote would have thrown all the affairs of this great Nation back into hopeless chaos." The President clarified his dilemma arguing that economic wellbeing was not fully restored and that the court stood in the way of a complete recovery. He then went on to rebuke the justices' previous decisions.

> When the Congress has sought to stabilize national agriculture, to improve the conditions of labor, to safeguard business against unfair competition, to protect our national resources, and in many other ways, to serve our clearly national needs, most of the Court has been assuming the power to pass on the wisdom of these acts of the Congress—and to approve or disapprove the public policy written into these laws.

Roosevelt stated that he designed his proposed legislation with no other purpose than to restore the balance between the three branches of government. As FDR explained to his listeners,

> Two of the horses are pulling in unison today; the third is not. Those who have intimated that the President of the United States is trying to drive that team, overlook the simple fact that the President, as Chief Executive, is himself one of the three horses.

Finally, he argued that the urgency of recovery made the realignment of power through a constitutional amendment impractical.

FDR bypassed Congress and went with what he perceived as improved odds. So, he sent the Judicial Reform Bill directly to the Senate. Despite the Democrats' healthy majority, the Senate remained unswayed by Roosevelt's arguments and voted 70 to 20 to let the bill die in committee. Some senators voted no because, although they believed Roosevelt's plan would improve the nation's economy, pragmatically they understood that eventually the tables would turn, and a Republican president might take advantage of the new law—what goes around comes around. However, the prevailing attitude among the senators was the need to reign in Roosevelt's power-grabbing tactics.

As it turned out, Roosevelt never had to submit the Judicial Reform bill. His presidential longevity solved that political problem as his three-plus terms allowed him to pack the court with hand-picked judges. Fortunately, President Roosevelt's progressive Supreme Court turned out not to be always left leaning. Albeit subtle, FDR's court-packing plan and the Schechter case may have instilled in present and future justices a greater appreciation of the US Constitution. In the decades following the Great Depression, the United States lagged Europe's trend to socialize government. A move slightly to the right, which was bad for future New Dealers, but wonderful for liberty and America's wealth.

## Jobs Programs

New Deal jobs programs presented another significant obstacle to economic prosperity. Due to lackluster results and the political shift to the right, they began to break down in 1938. Roosevelt's prized infrastructure and arts initiatives left enough valuable efforts behind to portray the illusion that the programs worked. In truth, their inefficiencies and cronyism placed a substantial toll on an economy struggling to recover. By the late 1930s, the reputation of New

Deal jobs programs struck bottom. A growing consensus considered them as charity organizations or political tools used to gain votes but not platforms for constructing valuable infrastructure. Leading up to the war, the Republican/conservative-Democrat coalition in Washington, frustrated by lackluster results and huge price tags, partially defunded FDR's jobs agencies. Others ended with the buildup and entry into World War II. The tremendous manpower needed for war challenged the argument for jobs programs.

A few initiatives remained intact to help the War effort, but by 1943, Congress finally shut down the New Deal's trademark employment agency, the Work Progress Administration. During its tenure, the WPA *employed* over ten million men and women. If programs like that had continued after the war, a full economic recovery would have been more difficult.

## Hyper Austerity

Any sinking economy induces some level of public frugality. Consumers, both at home and at work, spend less. They refrain from purchases that they consider luxuries. Resourcefulness rules the day. Survival toolboxes include staples such as chewing gum and chicken wire. Both men and women become self-sufficient with a needle and thread. The Great Depression marked a prolonged version of that type of penny-pinching culture. Then came the war, which knocked America's economic system further out of balance.

As a sacrifice for building an explosive military, from early 1942 until late 1945 much of America's civilian economy shut down. Economists' *unique* method of measuring economic growth, the GDP, masked those and other ensuing wartime economic problems. In truth, during the war, economic hardships intensified.

Besides the conversion of factories, the expansive war caused global commercial havoc. For example, the Japanese controlled much of the world's rubber supply. Freighters operating in Atlantic shipping lanes patrolled by the German navy ceased operations due to fear of U-boat torpedoes. Other trading nations had their own internal shortages to contend with, which reduced imports from and exports to the United States.

Between harmful government policies and the ravages of war, many consumer goods fell into one of three scarcity categories: unavailable, rationed, or out-of-stock due to market-based shortages. The following summary includes a small sampling of those goods that had become scarce. The federal government gave defense contractors priority for iron, steel, aluminum, nylon, fuel oils, and rubber. Per government decree, civilians could not buy automobiles, stoves, tires, and nylon stockings. A complex government-run rationing system limited purchases of sugar, butter, gasoline, and shoes. Almost everything else was in short supply: processed foods, cigarettes, bed sheets, milk, meats, typewriters, etc. Wartime spending did not revive an economy struggling to recover—it applied the brakes to it.

For economic slumps of all sizes, the greatest recovery tool is the potential created by prolonged austerity. In the case of the Great Depression and World War II, the frustration of minimalist living wound the recovery spring extremely tight. By 1947-ish, when normalcy finally arrived, it had been 17 years since Americans went on a lavish shopping spree. That is a long time to operate on a lean budget. Factories had been running with worn-out machinery. Few new housing starts meant multiple generations shared a home or an apartment for extended periods of time. Households and businesses were full of worn-out items, surviving on their last legs.

The long duration of the Great Depression, followed by wartime shortages, and finally the conversion back to peacetime, ramped up the demand to spend. Nearly 10 million GIs and 15 million defense-based factory workers were armed and ready for a post-war spending frenzy. Still, there was yet another tier driving the rush to splurge—innovation. Technology had made scores of consumer and industrial goods seem obsolete. Since 1929, manufactured products made significant advances, to the point that even if an individual, family, or business owned something that worked, the better *mouse trap* might be worth the purchase. For consumer-hungry Americans a surge was inevitable, but for how long?

# 45

# Peacetime Hangover

By the fourth-quarter of 1945, America again turned to recovery mode, primed for a solid economic reboot. But there were hurdles to overcome—short term and long term.

After four years of war, both businesses and workers had to adapt to peacetime conditions. Ten % of established companies failed to make the conversion from defense contractor to civilian supplier and closed their doors. GIs by the millions came home looking for decent work. Workers took pay cuts as they lost lucrative overtime pay when their factories converted from the production of military goods to the production of civilian goods. Others faced layoffs when businesses shut down to undo the defense-oriented retrofits. The new modern-era job market needed sorting out as the mismatch between workers' talents and the talents employers sought sent job hunters scurrying to learn new skills. These and other obstacles delayed the long-awaited recovery.

The end of the war did not immediately trip the recovery switch. Instead, immense wartime spending knocked supply and demand so far out of whack, that in many major industries the track toward a true recovery took a few years or more.

For example, Cincinnati Milling Machine, America's largest machine tool manufacturer for much of the twentieth century, built machines instrumental to the construction of planes, tanks, and other pieces of military equipment. After the war, the federal government sold thousands of used machine tools, flooding the market and driving down sales and profits to depression-era levels. The recovery produced enormous opportunities for the company but not until 1952, a delay of 7 years—see Figure 45.1.

Despite the post-war construction boom, Caterpillar, the leader in off-road construction equipment, suffered a dramatic decline in sales following the war— see Figure 45.2. Even though they had recouped the bulk of their losses by 1947, the construction equipment giant followed a trend similar to Cincinnati's; by the early 1950s, sales far surpassed wartime numbers.

American automakers, the heart and soul of the nation's economic engine, produced only a handful of automobiles in the four years prior to the Allies' victory. Potential car buyers anxiously waited for new models to arrive at their local dealerships. Compounding the difficulties of production ramp up were the continuation of wartime government price-control regulations, material shortages, and labor disturbances, which, including the sales needed to compensate for the previous four years, turned the industry's recovery into a five-year effort—see Figure 45.3.

Most of the obstacles were short-term issues, which could be resolved within a couple of years. Nevertheless, surviving New Deal labor legislation could not only delay the recovery, it had the potential to stifle the economic growth of future generations

## Organized Labor's Threat

As a critical component of the New Deal, Democrats sought to calm the perennial strife between labor unions and employers or anyone who got in their way. Federal labor laws had quelled much of the violence but as Figure 45.4 reveals, the number of strikes and lost man-days due to strikes skyrocketed. Organized labor disturbances, designed to shut down businesses in a show of defiance, created some of the largest post-war hurdles.

The Wagner Act granted organized labor tremendous power, but workers had to be patient with their newfound leverage. There was both timing and state-of-the-economy issues. Union leadership needed time to organize, then once in place, the economic woes of the Great Depression proved to be a daunting stage to flex labor muscle. Then came the war and Roosevelt's labor agreements. Beginning in 1945, even before world leaders had signed their peace treaties, labor unions smelled opportunity, and off came the gloves. Labor decided the best way to brandish their newfound strength was through massive coordinated general strikes. From 1945 into 1947 they walked away from their jobs in automobile factories, steel mills, mines, railroads, Hollywood studios, oil fields, and more. All totaled, workers protested by the hundreds of thousands.

Their stunt not only delayed the recovery, but their scheme backfired. The scale of the walkouts led a majority in Congress to realize that the Wagner Act gave organized labor too much power.

Leading the response to the labor uprisings, Republicans Senator Robert A. Taft (1889–1953) and Representative Fred A. Hartley, Jr.

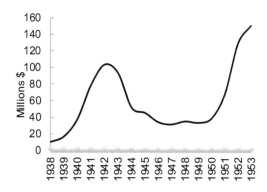

**Figure 45.1** Cincinnati Milling Machine sales, 1938-1953.

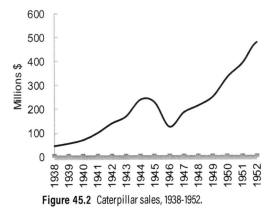

**Figure 45.2** Caterpillar sales, 1938-1952.

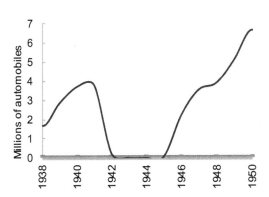

**Figure 45.3** Automobile sales, 1938 - 1950.

(1902–1969) introduced the Labor Management Relations Act of 1947, today known as the Taft-Hartley Act. Congress designed Taft-Hartley to weaken the 1935 Wagner Act by restoring some parity between employers and employees. Pro-labor President Truman vetoed the bill. In response, Congress and the Senate overruled the veto by a wide margin. Taft-Hartley became law on June 23, 1947.

The labor bill outlawed certain types of strikes and required union leadership to sign affidavits denying affiliations with the Communist party or any other organization declared an enemy of the state. Taft-Hartley prohibited labor unions from secondary boycotts, charging excessive dues, and forcing employers to pay for work not performed (featherbedding). The most significant Taft-Hartley regulation was its right-to-work clause which passed the authority for labor unions to force mandatory closed shops onto states instead of the federal government. After its passage, 26 of the 48 states voted to take advantage of the new labor law's wonderful gift to liberty and prosperity, restoring to workers their right to decide whether or not to join a labor union.

Even as Taft-Hartley reigned in labor's power, union membership continued to grow, enabling the disruption of 3,000 walkouts by 5,000,000 strikers annually for the next several years. Without the counter-labor union Taft-Hartley Act, those numbers would have been larger, to the point where America might have faced another economic disaster.

## Not so Small Government

By late 1945, with the war over and much of the New Deal abandoned, the recovery era was far from a Tea Party moment. Federal government *reductions* left a massive bureaucracy. For one, because of the Cold War, peacetime conditions did not mean peacetime spending. America had learned its lesson and continued to maintain a large and modern army, air force, and navy. America stockpiled armaments and manned military bases in such faraway places as Germany, Japan, Guam, and the Philippines. While military spending from 1947 through 1949 dropped to 8% of GDP, compared to almost 40% during the war years, the recovery-era armed forces cost ten times that of military spending during the late 1920s, which accounted for less than 3% of GDP.

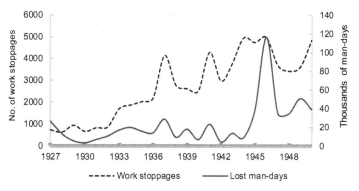

**Figure 45.4** Organized labor strife, 1927-1950.

Then there were the New Deal leftovers that gave Congress a novel way of expanding its influence—regulatory agencies. These mini governments within the federal domain could pass laws, spend money, and micromanage at a rate much faster than the several hundred members of Congress. The growing quantity of alphabet-soup agencies placed a heavy burden on a long-term recovery. From just a handful under the federal umbrella in the 1920s, thanks to the New Deal, that number flourished and then maintained that momentum into the 1940s. Post-war agencies included the:

- Federal Aviation Administration (FAA)
- Federal Communications Commission (FCC)
- Federal Housing Administration (FHA)
- National Labor Relations Board (NLRB)
- Securities and Exchange Commission (SEC)
- Social Security Administration (SSA)
- Tennessee Valley Authority (TVA)
- United States Housing Authority (USHA)

Besides the alphabet-soup agencies, massive departments such as the US Department of Agriculture and public welfare programs further weighed down the economy.

Then, although Congress scuttled the massive jobs programs during the war, they began work on its long-term replacement—the federal worker.

Higher paid, and owing to the nature of huge federal bureaucracies, much less likely to contribute to the economy, the tenured federal worker enjoyed great benefits and numerous opportunities for advancement.

Federal spending was a wash. It turned out, that as a percentage of GDP, post-war federal expenditures nearly matched the spending of the earlier isolationist-New Deal government. See Figure 45.5.

Big government demanded big taxes. To that aim, Washington maintained its heavy tax rate after the war, one that disproportionately taxed the wealthy, as the 'tax the rich' fervor did not subside in the post New Deal era. The top personal income rate of 91%, which was applied to incomes exceeding $200,000 seems ludicrous today. Even the lowest percentile at 20% feels high. Corporate tax rates ranged from 21 to 53%. The effective corporate rate averaged 35% when factoring in tax shelters. One of the cruelest taxes of all were the 'death' taxes, a leftover from the Hoover administration. In 1946, estates exceeding $50,000 were taxed at 50%. A gift tax, levied to prevent people from avoiding estate taxes, was 25%. The 1940s inherited the social security tax, but that at least remained at a mere 2%.

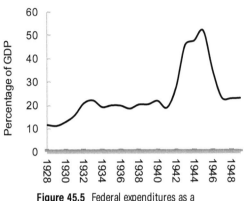

**Figure 45.5** Federal expenditures as a percentage of GDP, 1928-1950.

### New Technologies

What a mess—or so it seemed. On one hand, the American economy had to pick itself up all the while dragging along an enormous bureaucracy. Fortunately, within the American economy, covering hundreds of industries and many-thousands of products, the potential post-war scale of the private sector was massive. As a result, new-age technologies dramatically improved the life of almost every American. Modern capitalism's credits included improved labor-saving machinery, exciting job opportunities, world-class medical care, and more.

For the post-war recovery, unemployment rates dropped compared to the entire 1930s, while the aggregate volume of consumer, commercial, and industrial goods far exceeded those of the pre-panic years; however, the historic enormity of the era lies in the value of those goods. Economists lack the metric for a real comparison, but the degree of modernization suggests that working or not, Americans were noticeably poorer just a half generation earlier.

Once in high gear, the hyper-industrial machine bulldozed through the post-war government drag. Aside from a surge in defense spending during the Korean War (1950-1953), the shift to the right slowed down the rate of government expansion. What followed was an unprecedented period of economic growth that would last decades.

# Part Seven

# More Modern Marvels

# 46
# The Time Machine

As economist Paul Samuelson predicted the post-war recovery began with a struggle. Then, within a couple of years after the war's end the recovery stuck. The book closed on the Great Depression saga without an official end date or V-day-type celebrations.

Economists later calculated that from the Great Depression's start to finish, which included 11 years of panic, 4 years of war, and the recovery, American productivity doubled—at least according to their illusory monetary-based metric, the GDP. In reality, the GDP data was a huge understatement. Economists shortchanged America's industries by a wide margin.

Previous chapters chronicled the modernization process covering the early to midcentury years. Factories and mills during the 1930s produced modern marvels that ranged from innovative product designs to the development of advanced materials. From 1940 to 1945, the military-industrial system played its part, particularly in aerospace, plastics, electronics, and medicine. In the private sector, pockets of rogue designers whose factories lay idle because of the war made their modest contributions to modernization. Finally, a burst of industrial activity followed the war.

By the late 1940s, anyone not paying attention to new technologies, perhaps focused on surviving the 1930s, World War II, and the post-war's return to normalcy, might believe they stepped out of a time machine. Dramatic upgrades to modern marvels such as automobiles, locomotives, home appliances, construction equipment, buses, and even groceries appeared as if they all had come from the future. Those and many other products underwent major transformations compared to their 1920s counterparts. Some did not even exist at the onset of the Great Depression. Most impressive were their affordability and higher volumes. In classic American exceptionalism form, hundreds turned to thousands or thousands to millions. On the other hand, what defined modern technology? Media sources sent out constant reminders warning Americans not to get too comfortable with their latest gadgets as engineers and scientists seemed obsessed with making their latest inventions obsolete.

After the war, a slew of sci-fi-type technologies, some in their launch phase and others in late development, were about to hurl society into yet another era. Military advancements included nuclear power and jet propulsion. Bell Labs introduced the field of microelectronics with their 1947 patent of the transistor. Those game changers were so influential they had *ages* named after them. Television, the next big thing, became available in select markets. A nationwide sales explosion had to wait for cities to establish TV stations. Annual television sales broke the one-million barrier in 1949 and reached three million sets by 1950. This new communication medium garnered success for its wide range of content, complemented by the convenience of enjoying video broadcasts in the family living room. Not always well received, the boob tube's dubious value to society tarnished its reputation.

The combination of modernization and a revived economy did not end all suffering. Despite the wealth, many Americans lived in poverty. For some poor farmers, their livelihoods often came down to one of two bad choices: remain frugal and miss productivity-enhancing technology or invest in that technology and risk losing the family farm saddled with debt. Technological obsolescence continued to close factories and drive skilled trades to extinction. Corporations grew larger allowing a select few to blow apart the income gap. Slums and shanty towns declined, replaced by the social experiment of public housing projects. For many, prejudice and racism intensified their poverty.

✧✧✧

Industrialization's post-war burst brought a new round of social advancements, particularly the decline of deplorable racial discrimination, albeit at a painfully slow pace.

After the war, racial discrimination persisted as a national problem. Southeastern state politicians continued to enforce Jim Crow Laws. Much of the federal government practiced racial segregation. Even in the north, Levittown, the mega post-war housing tract on Long Island, New York, excluded minorities based on race or religion.

For the oppressed, enough was enough. The Fourteenth Amendment, which addresses civil rights and equal protection under the laws, had been in effect since 1868. When the economy returned to normalcy, early signs of the civil rights movement emerged. After all, 125,000 African Americans fought for their country overseas during World War II.

In July 1947, the Brooklyn Dodgers signed Jackie Robinson (1919-1972). Major League teams signed four more Negro League players that year. Baseball's new desegregation policy served as one of the few bright spots in an atmosphere of intolerable racism and segregation. In 1948, President Truman signed an executive order to abolish discrimination in the US Armed Forces.

At least the post-war society had begun the process of social reform. Still, it would be a long journey to catch up to the industrial renaissance.

✧✧✧

Mainstream economists and historians, oblivious to the robustness of the factory revival, struggled to understand the radical modern transformation. Smitten with implementing European-socialist ideologies, their histography of the Great Depression strayed far from reality. The conventional scholar marginalized manufacturing excellence, which was upstaged by Keynesian principals. After all, tracking money instead of commercial wealth required scholars to apply *advanced* mathematics. As for historians, they engineered a pair of brilliant maneuvers that heightened the value of left-leaning policies over the value of the goods and services generated by the private sector. First, they gave most of the credit for modernization to the government-sponsored industrial-military machine. Then, they credited the GI Bill for holding that momentum throughout the recovery.

These next chapters examine another batch of Great Depression technologies and how they matured and thrived after the war. This group again challenges academia's interpretation of the Great Depression with two notable takeaways. For one, real economic growth far exceeded the GDP's twofold increase. Second, big government contributed but played only a minor role in America's phenomenal newfound wealth.

# 47

# The Mold in Dr. Florey's Coat

Remarkable military-related technologies emerged from World War II, but one of the most valuable creations to come out of the war, and one of the greatest of the twentieth century, had nothing to do with weaponry. Led by the United States and Great Britain, its supporting cast included academia, research laboratories, nonprofits, government agencies, one of America's fastest-growing industries, and a generous amount of serendipity. The result was the mass-produced antibiotic drug, penicillin.

In 1928, Scottish biologist, Alexander Fleming (1881–1955), working for Saint Mary's Hospital in London, England, conducted a series of experiments searching for a practical antibiotic. Upon returning from holiday on September 28th, Dr. Fleming went down to his basement lab to check on a petri-dish containing a bacterial specimen. He found the dish uncovered, a careless mistake, but he noticed that a bluish-green mold had invaded the bacterial culture. A closer examination revealed, as the biologist noted, "a translucent ring surrounded the intruder." It turned out that the mold, a staphylococci fungus known as penicillium notatum, killed the bacterium on contact.

Encouraged by the discovery, Dr. Fleming turned his focus on extracting the chemicals from the antibiotic mold that attacked the bacteria. Fleming isolated the antibacterial source, which he named penicillin after the mold's Latin name, and in 1930 published a paper detailing his findings.

Students learn this fascinating bit of history in school, but that is just a small part of a much larger story. Before penicillin evolved into a wonder drug, its development required a massive collaborative effort, challenging deadlines, and a few more strokes of incredible luck.

Fleming believed penicillin might serve well as an antiseptic. Unlike existing germ-killing solutions of the time, Fleming's proposed drug was odorless and safe. But as for an antibiotic, the scientist doubted the drug could last long enough in the body to be an effective combatant against disease-type bacteria. Over the next couple of years, Dr. Fleming tried to extract a practical form of penicillin. Failing to produce a viable product, Fleming published his research and then shelved the experiment.

Fast forward 10 years from the original discovery of penicillin to 1938. A pathologist, Dr. Howard W. Florey (1898–1968), happened upon Alexander

Fleming's paper on penicillin. Florey headed up a team of scientists at Oxford University's Center of Pathology. The next year, in 1939, Germany launched major attacks against its European neighbors. It was a matter of time before Hitler's military turned its focus on England, and in response, the medical community anticipated a flood of wounded men and women.

For thousands of years it had been common for soldiers to survive a serious wound only to succumb to infection. The expectation of impending British war casualties heightened the interest in finding an antibacterial medicine. Florey believed that penicillin might be the key, and he had faith that his well-equipped lab and talented researchers could find a solution to turn penicillin into a game-changing medicine.

Dr. Florey obtained some of Fleming's cultures for testing. Within a year, Florey and his team extracted moderately pure penicillin in the form of a brown powder. In 1940, the Oxford team tested the drug on mice with positive results. The following year they tried penicillin on a human patient, a 48-year-old policeman with an infected facial cut. The only available dose halted the spread of the infection, but by doing so, they used the last of their supply and sadly the patient died within a month. Nevertheless, the team had proved the drug's potential.

Florey desperately sought financial backing to build a large-scale operation for penicillin production. By that time, England was fighting for its life against the Nazis, so funding became scarce. Penicillin research also had to compete with a class of drugs called sulfonamides, into which England had invested significant resources and funds and had developed much optimism for the drug's potential. Compared to modern penicillin, sulfonamides attacked a narrower range of bacteria and with less effectiveness. They did not kill the intended organisms, but only retarded their growth. Side effects included nausea, vomiting, and mental confusion. Worse, patients became hypersensitive to the drug after the first dose. Still, no antibiotic up to that time had received more attention than sulfonamide.

Meanwhile, Germany's air raids on England left labs and factories vulnerable. In response, he turned to his contacts in the United States for help, which in June 1941 still maintained its neutral status. That month, Dr. Florey received a well-needed grant from the Rockefeller Foundation. Soon after, Florey and team member Dr. Norman Heatley (1911–2004), a biochemist, made a daring escape from England to America. For the getaway, according to Eric Lax, author of *The Mold in Dr. Florey's Coat* (2005), Dr. Florey smeared his samples of penicillin onto his coat to prevent the drug from landing in German hands if he were to be captured. Once in America, the scientists set up shop in Peoria, Illinois. Peoria had a new US Department of Agriculture facility, the Northern Regional Research Laboratory, dedicated to the development of medicines derived from corn waste—specifically, byproducts from corn milling.

At the Peoria lab, a microbiologist, Dr. Charles Thom (1872–1956), had developed a carrier agent from corn-steep liquor designed to deliver medicines into the human body. The equipment used in Thom's lab resembled that of a brewery's. Experiments showed that when applied to penicillin, the medium increased the potency of the drug by tenfold.

Despite advances, the team needed more funding to help find a viable commercial solution to produce penicillin. At first, nobody in the United States government seemed interested; American researchers believed they were on the verge of a breakthrough with sulfonamide-based drugs. After the attack on Pearl Harbor, American interest in penicillin intensified, so with the help of the US Department of Agriculture, the military declared the development of mass-produced penicillin a war project.

High-volume production of penicillin proved difficult. After a year of research, Peoria's only source of penicillin came from the descendants of Fleming's original culture. The first patient to survive a deadly infection thanks to penicillin required one-half of the world's supply. So precious was the drug, scientists kept the supply up by recycling a patient's urine. Desperate for a better solution, investigators conducted a worldwide search for a more manufacturing-friendly strain. The hunt went full circle around the world and ended back at Peoria. A scientist at the Institute, Mary Hunt, or as the English referred to her, Mouldy Mary, stumbled upon a cantaloupe with the telltale bluish-green mold while shopping for groceries. The new host turned out to be the discovery scientists needed. Workers at the lab soon found ways to extract and process the drug 200 times faster than Fleming's sample. With the help of ultraviolet light (UV) and X-rays, scientists mutated the mold to further enhance the drug's manufacturing capabilities. Finally, they achieved sufficient scientific advancements to build a high-production factory based on Peoria's latest deep-tank fermentation methods which included just the right amount of aeration and agitation—time for the next stage.

Besides mammoth technical obstacles, the mass production of penicillin also faced the overbearing bureaucracy of the United States government. Roosevelt despised big business, and he routinely unleashed his goons from the Department of Justice to harass successful commercial enterprises. Their favorite weapon, the Sherman Act, imposed felony charges on its worst offenders. Convictions could result in jail time and massive fines. The drug makers needed to be extra careful with penicillin. As with other industries, the complex technology of pharmaceuticals demanded a coalition of specialists to patent and manufacture their products. Those partnerships required deal making, exclusions, and elaborate secrets. Even though research and new-product launches placed a heavy burden on companies, if successful, they would produce huge profits. Launching new drugs, like all products, needed to follow the laws of supply and demand, which meant that the price for drugs often started out very expensive before becoming affordable. All those actions, from the patent process to the pricing structure, could violate antitrust laws. The drug industry at the time consisted of numerous small firms. A windfall innovation product such as penicillin could place a company in the cross hairs of FDR's DOJ thugs.

The drug companies hesitated but an approval nod from the DOJ helped make the decision for drug companies to move forward. Merck and Company, formerly part of a German corporation expropriated during World War I, took the lead with Charles Pfizer, ER Squibb, and Abbott Laboratories playing supporting roles. All totaled, the drug companies built 21 plants dedicated to manufacturing penicillin in bulk.

The nation's first female chemical engineer, Margaret Hutchinson Rousseau (1910–2000), headed the project at Merck. She used deep-tank fermentation with a recipe of corn-steep liquor, penicillin powder, and other ingredients. Combined, the pharmaceuticals' processes produced astonishing results. Soon after the start of production, Merck's New Jersey facility sent a 32-liter supply of penicillin to treat burn victims of the November 28, 1942 Coconut Grove fire in Boston which killed 492 people. That next spring, the drug firms provided penicillin for two army hospitals, and by June 1943, a few combat zones began receiving the valuable drug.

By June 1944 scientific discovery had turned to mass production, American stockpiles of penicillin reached 2.3 million doses, just in time for the Allied Forces' D-Day invasion of Europe. Penicillin shipments to France saved thousands of lives, mostly by healing infections from wounds and amputations. The next year, the drug industry produced four times that amount with the majority still going to wounded soldiers, but with an ever-increasing volume allocated for civilians. The cost of the wonder drug plummeted with volume. With an exorbitant price in 1942, it dropped to $20 per dose in 1944, and then down to 55 cents in 1945. Through high demand, a considerable volume, and a low price, penicillin became pharmaceutical's first high-production drug. It saved hundreds of thousands of lives over the next few years and many millions over the following decades.

For their extraordinary work on penicillin, Fleming, Florey, and Ernst

B. Chain (1906–1979) shared the 1945 Nobel Prize in Physiology or Medicine: Fleming for discovery, Florey for commercialization, and Chain for deciphering penicillin's structure. Dr. Thom won some important patents, but his reputation took a hit when he chose, for his own maximum financial gain, not to credit Dr. Heatley's work on critical patents.

The income, knowledge, and confidence gained from the penicillin venture led to the development of more drugs, including antibiotics. Researchers also continued to improve penicillin. One development increased its efficiency allowing a body to require fewer units. Then in 1958, the pharmaceutical companies, after extensive research, introduced a synthetic version of penicillin.

The penicillin saga has several interesting elements to it. Besides being a remarkable story, there was the tremendous amount of cooperation required. The resources responsible spanned two nations and included: a lone researcher in a small hospital, a research team in a private college, funds provided by a government agency, funding from a charity founded by an American robber baron, a government research lab, the United States military, and private corporations. The breakthrough development also highlighted the importance of continuous improvement that typically occurs after an invention's introduction. Finally, there was the matter of drama, credit, and humility. Fleming, by far, has received the most credit for his role as the spark. Dr. Heatley did not receive a Nobel prize, as the rules limited the committee to select only three recipients. Nor was it just that a federal employee (Dr. Thom) hijacked critical patents. Life is not always fair, and neither were the financial rewards and credit received for all those involved, but the three primary researchers Florey, Chain, and Heatley rose above petty egos, content with their contributions to humankind.

# 48

# The End of the Iconic Steam Engine

Railroads served an integral part in America's nineteenth-century journey toward modernization. Their role as a high-volume/high-speed method of transportation helped transform the nation from its colonial, agricultural economy into the world's greatest industrial powerhouse. Rail's contribution to commerce made the business of railroads the wealthiest in the country. Then, 100 years after the first commercial line began operation, and still within sight of its heyday, the mighty enterprise became embroiled in a fight for survival.

Already in trouble because of government overregulation, railroads faced further financial hardships as the Great Depression intensified. By 1930, the economic panic threatened to place many in the railroad industry into bankruptcy, and most of the rest into debt. However, it was the rivalry of the versatile internal combustion engine that created a longer-term threat.

The passenger side of the railroad business became the first casualty. In 1920, riders made more than a billion trips on steam-driven trains but by 1933, that number plummeted to less than half a million. Buses were one of rail's early rivals. Several million registered automobiles, with access to the nation's developing network of highways, established themselves as an even greater threat to passenger train service. The automobile also encouraged population growth into the suburbs and away from railways. The much faster commercial airplane, although still in its infancy, presented yet another viable alternative to rail travel.

Fortunately, in 1930 the rail outlook for hauling freight appeared more promising. Railroads infrastructure included over 250,000 miles of track. Trucks, although competition, also proved a valuable asset moving goods back and forth between rail terminals and customers.

The railroad industry believed they could survive, but they needed to adopt an aggressive strategy of continuous improvement. In response, the railroads launched a multi-pronged effort to modernize, particularly in locomotive propulsion. The endeavor included improving the efficiency of steam power plus adding propulsion alternatives electric motors and diesel engines where practical. Since railroads were still big business, locomotive manufacturers, despite the obstacles of a depressed industry, continued to deliver large amounts of rolling stock during the 1930s. This provided enough income for generous research funding.

Progress during the Great Depression included developing larger engines. Since bigger is often more efficient, manufacturers built the largest steam engines ever produced. To further power those larger machines, the industry increased the steam pressures, going from 150 pounds per square inch (psi) in 1895 to between 250 and 400 psi in the 1930s. By the early 1940s, they operated at 500 psi.

These new locomotive workhorses, weighing more than 550 tons each, came equipped with a massive steam engine capable of producing 7,000 horsepower. These machines were not just bigger. Designers implemented improvements that lowered soot emissions, improved safety, and increased fuel efficiency. Additional upgrades reduced maintenance costs. One option that broke tradition added coal tender conveyors that replaced the shovel to help the firemen move coal.

More efficient running gear and modified wheel arrangements augmented the larger engines. Those new classes of locomotives pulled heavier loads at higher speeds, some approaching 110 miles per hour, thus reducing freight costs. Meanwhile, as steam-engine development continued, two other power technologies, the veteran electric motor and the emerging diesel engine, proved promising for the future of the railroad industry.

✧✧✧

Following in the footsteps of Thomas Edison's earliest electrical grid, rail entrepreneurs of the late 1880s developed a new market, the trolley car, aka streetcar. Trolleys lifted city travelers from the oft-muddied streets and delivered them rapidly and conveniently to their destinations. Powered by electric motors, lighter-duty trolleys combined the locomotive and bus into one vehicle. Electrical energy was transferred from overhead via the familiar trolley pole. Their rail lines occupied a combination of dedicated rail rights-of-ways and city streets, the latter which they often shared with other types of traffic.

Trolley cars featured numerous advantages over steam, particularly in urban environments. They were less expensive to manufacture and cheaper to operate, accelerated and decelerated quicker, drove both forward and backward, and ran pollution free—as opposed to the noxious soot and smoke generated by steam locomotives.

Streetcars followed the spread of the nation's rapidly expanding electrical grid. By the 1920s, virtually every city and major town in the United States had a trolley car system. Not just constrained to Main Street, trolley companies built interurban lines that spread out across the nation, some stretching over

200 miles. The faster trolley cars were capable of speeds of 80 miles per hour. At their peak in the early 1920s, more than 1000 companies transported 14 billion passengers a year.

Through the 1930s and 1940s, despite fierce competition from buses and automobiles, trolley manufacturers continued to strive for improvement, especially in the areas of safety and comfort. During that time, a consortium of electric-rail people, the Electric Railway President's Conference Committee, began work on an advanced streetcar. Now known as the PCC (named for the design committee), the redesigned streetcar, with its sleek art deco motif, more resembled a bus than a traditional trolley. Design features included doors located on the curbside, front, and middle, which made entering and exiting convenient. A large ventilation system and electric heaters kept riders reasonably comfortable. The PCC came equipped with a triple braking system—each one designed for a specific mode. For planned stops, trains used dynamic-regenetive braking, which conserved energy, but for emergency stopping, the trolley relied on friction air brakes. Powerful magnetics secured the parking brakes.

First released in 1936, the PCC experienced rare success for a designed-by-committee machine. Before production ended in 1952, approximately 5,000 PCC trolleys were built in America, and possibly 25,000 worldwide. Prewar models continued to run into the 1980s. Several cities operated World War II era PCC trolleys as historic streetcars. Boston's Massachusetts Bay Transit Authority (MBTA) operated a fleet of refurbished PCC cars for regular service on portions of their Ashmont-Mattapan High Speed Line into the twenty-first century, although nostalgic PCC street cars traveled at lower speeds.

✿✿✿

While steam and electric propulsion became well established by 1930, another faction of the railroad industry pursued a revolutionary locomotive propulsion system, the diesel engine. Despite the lower price of coal, fuel costs ran about three times higher for steam compared to diesel. Steam engines required a lengthy process for both startup and shutdown of the engine and boilers; diesels needed much less. Unlike the diesel version, steam locomotives required frequent maintenance, a water supply for its boiler, and produced significant soot pollution. Steam power also incurred the extra labor cost of delivering coal to the boilers. Mineral deposits played havoc on boiler systems. The made-to-order nature of locomotives, depending on loads and grades, prevented standardization. A steam locomotive's complex linkages involved frequent lubrication along with the associated costs of downtime. Customization made stocking a complete inventory of critical spare parts almost impossible, while replacements of limited-production parts could take weeks. Steam-locomotive industry leader Baldwin at one time offered nearly 500 different models.

The earliest diesel locomotives were smaller engines without electrical generation. These found work in rail yards. Called switchers, they moved cars to and from trains. Diesel switching locomotives gained favor over steam engines by the mid-1930s, making the yards a valuable proving ground for diesel engines.

In 1915, industrial giants including General Electric, Ingersoll Rand, and General Motors launched a long-term developmental program to build a practical diesel-electric locomotive. Their design used a shipboard diesel engine and an electric generator coupled to an electric-powered traction drive like those used on trolley cars.

Diesel-electrics had some major advantages over diesel-only designs. Any internal-combustion engine lacks the capability to operate at slow speed unless coupled with a complicated clutch and transmission system, which for massive loads appeared prohibitive. Diesels with a generator-motor drive system allowed the locomotive to operate at low speeds. Another significant advantage was that electric motors allowed for four drive wheels that eliminated the classic configuration of complex linkages. Multiple driven wheels also improved a locomotive's tractive forces.

Diesel-electric builders first chose to revolutionize passenger service. The Edward G. Budd Company introduced its Pioneer Zephyr in April 1934. With an appearance unlike any steam locomotive of the time, the stainless steel-clad Zephyr became one of a class of locomotives known as streamliners, and for obvious reasons. Sleek and aerodynamic, diesel-electric streamliners operated with three or four articulated (coupled) passenger cars.

The Zephyr was the fastest train built up to that time. Its top speed exceeded 110 miles per hour. A month after its delivery to the Chicago Burlington & Quincy Railroad, the Zephyr covered a route from Denver to Chicago, 1,015 miles, in 13 hours and 4 minutes, an average speed of 76½ mph. Numerous manufacturers built various versions of streamliners, all with articulated passenger cars. Designed not only for speed but also comfort, streamliners featured air-conditioned cars earlier than air conditioning became available in automobiles.

The Santa Fe Railroad debuted the Super Chief, a larger streamliner, in 1937. A 3,600-horsepower diesel engine powered the sleek General Motors locomotive. At 2.5 times the cost of steam-powered trains, the diesel train featured luxury and speed. Its lightweight aluminum-bodied Pullman cars came with air conditioning in the private berthing rooms, dining cars, and parlor cars. The Super Chief cruised the western states at 80 to 90 miles per hour, and compared to steam-driven trains, there were fewer stops and no switching of trains due to varying terrain.

The popularity of streamliners, as with all passenger trains, plummeted after the war. Plus, railroads opted for the flexibility of coupled cars, however the diesel engine's advantages over steam engines for pulling loaded cars remained. While diesel electrics found a niche as rail yard locomotives and rapid-transit streamliners, steam engines still owned the more lucrative freight market—for the moment.

After 1930, few passenger rails earned a profit. The exception was the 1942 to 1945 World War II years. At the time, automobile manufacturers had converted to wartime production, and rationed items such as tires and gasoline became scarce. Freight, on the other hand, remained viable and a long-shot candidate for diesel-electrics.

Compared to the massive steam engines of the early 1940s, complicated diesel-electrics were more expensive and less powerful. The high capital cost of an engine and drivetrain remained a major drawback that discouraged railroads from purchasing diesels instead of steam engines. Also, diesel-electrics required less precision and fewer parts. Another timeless roadblock to change was tradition. Railroads had substantial capital invested in steam engines, along with the knowledge and facilities to repair and maintain them.

Eventually, locomotive builders tapped the diesel's potential leading to lower prices and provide sufficient power for hauling heavy loads. The control an operating engineer had over the power output and speed of a diesel engine compared to a steam engine became the deciding difference between the two engines. A single engineer in a lead diesel locomotive could control multiple connected engines. Steam locomotives running just two engines, known as double heading, required a team of operators for each locomotive, yet still presented technical challenges, thus the railroads usually avoided the practice. As a result, railroads during the steam age ordered locomotives to fit their applications—those with lighter loads ordered smaller engines, and those with heavy-duty requirements ordered larger engines.

Because railroaders could link an entire line of diesel-electrics—two, three, four, or more depending on the load, with little more operation effort than that of a single locomotive, manufacturers offered a limited number of standard sizes. Consequently, a series of 2,000-horsepower diesel-electrics could perform the duties of a single 4,000 or 6,000 horsepower steam engine. Standardization encouraged higher volumes, which led to practical mass production principles, and in the process closed the cost gap on the specialty craft production required

GM 1947 Electromotive F-Series Astra Liner diesel-electric locomotive. *Courtesy General Motors Archives*

for steam locomotives. By combining modern technology and economies of scale, diesel-electric manufacturers turned the cost tables on traditional steam locomotives. Based on the ability to use a team of engines, locomotive builders *rationalized* their product line to meet market demands with a leaner selection of easier-to-build smaller locomotives.

For further cost savings, the nature of diesel-electric locomotives lent themselves to a more modular, integrated design. Unlike steam engine builders, diesel locomotive manufacturers designed diesel electrics similar to trucks. Their three main assemblies consisted of a chassis, suspension, and wheels; engine, drivetrain; and carbody plus interior. This mindset allowed train manufacturers to pursue a horizontal supply chain.

Various companies took responsibility for those main components. For example, an Alco locomotive might come equipped with a General Electric drive and motors driven by a Fairbanks Morse diesel engine. A company skilled in state-of-the-art sheet metal work would supply the body. Another vendor proficient in welding, forming, and machining technology built the chassis. Once again, the longstanding practice of division of labor methods produced a quality product at a lower price.

When set into practice, railroaders realized further advantages with diesel electrics. The diesel-electric's reduced weight meant it accelerated quicker, and its lower center of gravity improved handling around curves reducing rail wear. Sleek aerodynamics saved fuel at higher speeds. By the mid-1930s, diesels only served short-distance passenger service. Through the 1940s they conquered freight and long-distance passenger service. Unfortunately for those passionate for the nostalgia of steam locomotives chugging along the tracks, steam's time had passed. Despite the railroads' best efforts, the last steam locomotives produced in America rolled out of factories in the mid-1940s—signaling the end for steam locomotion. The major lines had replaced all their steam engines with diesels by the late 1950s. On the bright side, diesel-electrics killed the steam engine, but they saved the railroad industry.

The ability to haul raw materials from distant places into a centralized location and then deliver finished goods across the country to consumer outlets made railroads a critical component of the Industrial Revolution. A larger post-war rejuvenated economy needed railroads to up their game, which they did. In 1950, annual railroad freight in ton-miles exceeded 1930 amounts by 50%.

# 49

# Material Movers from Earth to Factory Floors

Engineering marvels such as track drives, the diesel engine, hydraulic controls, and steel weldments all came together during the Great Depression to produce modern construction equipment. The ability of those massive machines to move prodigious amounts of earth made them invaluable for the war effort and critical for the post-war construction boom. They plowed through German fortifications and built dams, airfields, roads, canals, and strip mines.

In 1904, the Holt Manufacturing Company out of Stockton, California, demonstrated unique continuous crawler treads with individual left and right controls. Designed for agriculture, the steel tracks became most famous when adopted for World War I tanks and later for construction equipment. The company merged with a competitor in 1925 and changed its name to the Caterpillar Tractor Company. Its crawler treads became known as caterpillar tracks.

Columbia Steel Company at Geneva, Utah. Bulldozer used in grading during the construction of a new steel mill which will make important additions to the vast amount of steel needed for the war effort-1942. *Courtesy Farm Security Administration-Office of War Information*

Caterpillar designed their machines for pulling heavy loads over sloppy ground where wheeled vehicles would lose traction. By the 1930s, the addition of a heavy blade, supplied by third parties, converted a tractor into a bulldozer. After the war, Caterpillar offered a variety of their own dozers from small to mammoth. Other enhancements combined to make the modern bulldozer *the* formidable earth mover.

Caterpillar became interested in diesels during the late 1920s. In 1931, they introduced their first tractor diesel engine. By 1933, diesel orders surpassed gasoline engine sales as Caterpillar became the largest manufacturer of diesel engines. The higher efficiency and natural robustness of diesel engines made them the ideal choice to power off-road equipment. Unlike their use in some transportation vehicles, a diesel engine's heavy weight was an advantage in construction equipment.

Fluid power technology, in this case hydraulics, formed another key field that modernized construction equipment. Hydraulic power had already revolutionized manufacturing, aviation, military hardware, and automobile technologies during the 1920s and 1930s. By the mid-1940s, construction vehicles had a large variety of well-designed parts such as pumps, valves, actuators, fittings, and hoses, all of which were designed to lower, lift, or articulate a variety of devices and attachments.

Excavators, with early versions known as steam shovels, are one the earliest heavy pieces of modern construction equipment. They shoveled large swaths of hard-to-reach earth. Early steam shovels rode on rails similar to locomotive tracks. After the war, although sometimes still referred to as steam shovels, excavators had upgraded to diesel engines. Caterpillar tracks replaced wheels. Hydraulic power actuated their massive mechanical arms instead of winches and cables. Excavators came in a variety of sizes providing the right tool for the job.

Caterpillar developed another distinctive type of construction equipment for smoothing out roads—the grader. The construction giant abandoned tracks for tires, built a frame on an extended wheelbase, and strategically placed the plow blade in the center of the vehicle. Graders became a valuable tool during the post-war era of prodigious road building.

✿✿✿

Laying sidewalks with grader at New York World's Fair-1939. *Courtesy New York Public Library*

While manufacturers designed one type of equipment for off-road conditions, a sister industry developed forklift trucks and tow motors for the factory floor and commercial yard work. Instead of mighty diesel engines, these workhorses, for indoor safety, used either a propane-fueled gas engine or a battery-powered electric motor. Larger machines, or those with high-duty cycles, relied on propane-fueled engines, while smaller units that were used intermittently adopted electric power. A tow motor is a tractor designed for smooth conditions. They are compact but heavy with minimal ground clearance to lower their center of gravity. For a conversion to forklift the tractor adds a vertical, hydraulic lift with adjustable forks.

Forklift trucks and tow motors are two more machines introduced in the 1920s, modernized during the 1930s, and becoming prolific by the 1940s. Those versatile trucks tackled material handling tasks with significantly greater efficiency and spanned nearly every business sector. In the case of fork trucks, warehousing went vertical, reducing storage square footage, thus saving capital and cutting transport time.

# 50
# I Should Just Go Home and Ride My Tractor

Henry Ford's Fordson tractor revolutionized farming in the 1920s; however, the new and improved generation of tractors available to farmers in the 1940s far surpassed the machines from 20 years past. The Ford crew, in their early attempts to build an affordable tractor, struggled through a huge learning curve. The tractors were too expensive for the average farmer, unreliable, sometimes dangerous (prone to tipping over), and lacked basic functions found on more modern machines. Then came the Great Depression. Profitable farming proved difficult and so did selling tractors to struggling farmers. Suppressed prices of farm goods made large purchases nearly impossible. For many, horses would have to do for several more years. Farm tractor sales dropped 90% over the first two years of the panic, while the number of tractor manufacturers fell from 60 to 9. The survivors, as with so many other industries, worked hard to advance their industry.

1946 catalog illustration of an International Harvester Farmall HV tractor. *Courtesy Wisconsin Historical Society*

There were many design upgrades, but 1940s tractors shared basic concepts pioneered by the Fordson tractor, including a heavy cast iron backbone that housed the engine, transmission, and differential gears. The driver sat propped up between two large traction wheels with the smaller front wheels providing steering.

For improved reliability, manufacturers had access to higher-strength steels, which made a tractor's engine and the other drivetrain components including gears, bearings, shafts, etc., more robust. As a result, with nominal vehicle maintenance, tractors could perform their farm-grade duties for many years.

Rubber tires replaced traditional steel wheels in the early 1930s. Rubber tires sacrificed a bit of traction but improved fuel economy. They did not tear up the earth as much and rode smoother than steel. A softer ride saved wear and tear on both the operator and the equipment. Plus, rubber tires hauled loads at higher speeds—a major productivity booster. By the late 1940s, most older tractors had their all-steel wheels converted to rubber tires.

Another innovation, dating back to the early 1930s, allowed farmers to work on rows of crops. The row crop-designed tractors had a tricycle configuration. Two large rear wheels allowed the rear axle to straddle and clear crops, while the two front smaller wheels, mounted side by side, rode between the rows. Some tractors had adjustable spacing of the front wheels, so they could be configured for row crop or standard operation.

The three-point hitch (named for the arrangement of linkages) vastly increased the utility of a farm tractor. When assisted with oil-pressurized cylinders (hydraulics), the detaching and attaching of heavy farm implements became a one-man job. Those implements connected to either the front or the back of the tractor. Some tools towed by a tractor were self-propelled. The combination of many tractor accessories aided the farming of nearly every crop and included all the tasks from planting to harvesting. There were other conveniences. For easy starting, an electric motor replaced the hand crank. The PTO (power take off), from the 1920s, gave the farmer added flexibility and was safer thanks to an independent clutch.

After the war, tractor sales picked up, and the industry regained its momentum to revolutionize farming. Manufactures now offered a wider variety of sizes, although the regular size fit many applications. For bigger farms there were larger tractors, and for the homeowner and hobby farmer, miniature versions.

To the extent that technology modernized farming during the Great Depression consider that in 1930 American farmers operated 1 million tractors—by the end of the 1940s, 3.5 million. Besides their tremendous utilitarian value, tractors had a big-boy's toy element to them. As the Republican Senator from Iowa, Chuck Grassley (b. 1933) once quipped, "Maybe I should just go home and ride my tractor."

## 51

# Detroit Steel Back on Top

Automakers turned out a staggering number of vehicles leading up to and through the Great Depression. In one generation, between 1920 and 1942, Detroit built 60 million automobiles. By 1945, half of them remained registered. To keep car factories humming at that blistering pace required an enormous amount of raw materials, workers, subcontractors, machinery, and research. Outside the factory, an extensive service network kept American wheels turning. Just as extraordinary was the major role that the automobile played in modern American society.

Following the war, logistic difficulties prevented Detroit from picking up where they left off in 1942. Besides the massive retooling effort, material shortages and labor strikes besieged the industry. Carmakers released recycled designs of their 1942 automobile models at subpar 1930s sales rates for 1946 and 1947. Meanwhile, engineering departments continued to work overtime on new models. Desperate to stay ahead of the competition, and with the advantage of a slew of recent innovations, auto companies sought to revolutionize automotive design. The effort paid off. Record sales by the end of the decade drove the post-war auto industry to regain its title as the heartbeat of the American economy and assured a long and fruitful recovery. A few years later in 1953, Charles Wilson (1890-1961), the president of General Motors, famously declared something to the effect that: "As GM goes, so does the country."

Studebaker became the first car builder able to break from the pack. Smallness allowed the automaker to be nimble and avoid the wrath of labor unions. Studebaker released their newly designed Commander and similar models in 1947. Despite the car's dramatic styling, annual sales ran only about 15,000 units for model years 1947 through 1949.

The avant-garde design team of the new modern era came from Ford. General Motors' Chevrolet dominated car sales from much of the mid-1920s on, but the Ford brand led the field in sales leading up to the Great Depression and did the same following the economic panic. The first triumph began when, in 1927, Henry Ford's managers convinced him to move forward from the Model T. Begrudgingly he shut down his entire automotive operation for nine months to retrofit his factories for the next-generation automobile. Ford introduced their larger, more powerful, updated version of the Model T in 1928—the

Model A. While advanced compared to its competitors, the Model A renewed the boxy outline, attached fenders, exposed radiator, and flat windshield styling of the Tin Lizzy. Other Model A features included spoked wheels, a four-cylinder engine, a non-synchro three-speed manual transmission, and no power anything.

Ford once again bested the competition shortly after the war with its 1949 Ford Custom. The new Ford models, such as the 2-door coupe, shown on this page, stand out as dramatic upgrades from its pre-war models, and especially compared to the Model A. The radical designs forced Ford's competitors back to the drawing board, evidenced by their subsequent release of similar designs.

1929 Ford Model A roadster. *Courtesy The Henry Ford*

Style played a large part in the love affair between Americans and their automobiles. To accommodate the latest fashion trends, automakers continued to revise each car's style. The Model A shown, which could have been purchased close to Black Tuesday, had moved well past the horseless carriage look but still sported a squared-off, Victorian motif. For the latter 1930s, designers smoothed the edges, sloped the windward vertical surfaces to improve aerodynamics, and added some art deco influences (see Chapter 33). The big changes came after the war. Ford released its 1949 models in June 1948 during the outer limits of the Great Depression. For their jet-age inspired design, Ford dropped the appendages and integrated the fenders, radiator, lights, bumpers, and door hinges. The automaker expanded the passenger compartment and added trunk space by adopting a cab-forward design. To achieve that radical look, Ford moved the engine over the front wheels and positioned the back seats in front of the rear axle. As a result, the new Ford sat four inches lower than its predecessor.

The exterior design changes only begin to hint at the improvements made. Two innovations allowed for the modifications that lowered the car's frame and body. First, a floating driveshaft, with U-joints that ran through a tunnel integral to the frame, replaced the clunky, decades-old torque tube and Hotchkiss drive arrangement. Second, hypoid gears, used to transmit power from the driveshaft to the rear wheels, helped lower and reduce the size of the differential and rear axle.

Improved front and rear suspension softened the ride yet maintained firmness when taking sharp corners, which when combined with the lower center of gravity further enhanced performance. The 1949 Ford lacked a modern unibody structure; however, better integration between the frame and the body along with heavier welds brought the car, built at Ford's mega-sized River Rouge complex, one step closer. Compared to its equivalent 1948 model, for the 1949 Ford, engineers shed 250 pounds and increased the car's cornering rigidity by 25%. Optional hydraulic power-assist, along with improved kinematics in the steering linkage, improved steering response. There were also larger drum brakes, which when coupled with hydraulic power assist, reduced the stopping distance, and improved brake reliability. Another safety feature, turn signals, came standard. For added comfort, car buyers could purchase dealer-installed air conditioning, and for entertainment, an in-dash AM radio.

1949 Ford Tudor sedan. *Courtesy The Henry Ford*

Underneath the art were refinements that not only improved performance, handling, drivability, and reliability, but importantly safety. Today, those late-1940s cars take a lot of criticism for being death traps. It is interesting how 70 years of hindsight heightens perception. The automobile's larger safety problem then is the same one today—Americans are willing to forgo safe driving in the place of adding thrill and impatience to the sport of driving.

Post-war drivers got to experience one of the automobile's companion technologies, the limited-access highway. In 1940, the Pennsylvania Turnpike opened. As the modern road spread, it would push the engineering of the automobile and encourage an interstate highway system allowing more Americans to explore the USA by car.

Today, the US Department of Transportation tracks an interesting statistic—the average age of registered motor vehicles. Their figures challenge the idiom "They don't make them like they used to." For automobiles, the data shows an upward-sloping curve from a low of 5.1 years in 1969 to a still rising 11.5 years in 2016. Unfortunately, similar stats prior to 1969 are unavailable, but if circa 1950 Hollywood street scenes are an indication, the stylish new post-war-recovery models sped the journey of clunkers to the scrap yard.

# 52

# From Shops to Labs

As product designs and manufacturing methods became more complex, then so did the inventive process. As a result, more innovation took place in corporate laboratories. However, there were a few notable exceptions of inventors during the 1930s and 1940s who succeeded mostly on their own.

The inventor Edwin Armstrong (1890–1954) should be a more recognizable name outside of those knowledgeable in the telecommunications field. Armstrong spent most of his career working at Columbia University and RCA, but much of his work came from his independent and prolific inventing that spanned over 20 years, from 1913 into the 1930s. Armstrong's early legacy is that he contributed as much to the fledgling electronics industry as any other inventor. His development of two critical electronic devices, the regenerative circuit and super heterodyne, dramatically reduced static noise and improved the tuning of amplified sound. Armstrong's inventions improved the quality of talk and music through a radio's loudspeaker. Gone was the faint crackling sound through the requisite headset.

Armstrong's most famous invention was static-free FM (frequency modulation) radio, which he patented in 1933. Unfortunately, his legacy is more about legal battles than engineering prowess. He spent an inordinate amount of time, particularly the 1940s, in the courts defending his ownership of intellectual property. Big corporations, such as RCA and AT&T, along with a few competing inventors, used the courts to deny him credit and royalties for his early inventions and later FM radio. The future success of FM radio also ran into a couple of obstacles from the Federal Communications Commission (FCC). First, they were influenced by RCA who had their internal battles with Armstrong, and second, technical controversies arose within the federal agency regarding the operation of frequency bands used for FM radio.

Broke and destitute, Edwin Armstrong committed suicide in 1954. After his death, Mrs. Armstrong continued to battle for her husband's patent recognition. Higher courts reversed many of the earlier decisions that had gone against Armstrong, which made her a wealthy widow. Edwin Armstrong's inventions were ahead of his time. While the controversies at the FCC delayed the FM radio, industry adopted the sound format for television. The development of stereophonics finally made FM radio commercially viable by the 1960s.

Another innovator of the era whose fame completely eludes celebrity status was the Danish-American engineer Niels Christensen (1865–1952). Christensen pioneered several inventions while moonlighting in his basement. His most notable was a simple fluid seal named an O-ring. Inventors had tinkered with O-ring concepts without success since the turn of the century. Christensen's success depended on two breakthroughs. First, he discovered some new-age polymers that worked well for both the performance and manufacturing efficiency of his seal. Second, Christensen found the optimum O-ring groove proportions that gave the seal its ideal squeeze. He received a patent for his invention in 1939.

An assortment of O-rings. *Courtesy Grayhawk Press*

Christensen targeted his O-ring to address a weakness in the growing fluid-power market. Fluid power, using either compressed air or pressurized hydraulic oil, had become a valuable tool in many industries such as machine tools, automobiles, aircraft, construction equipment, and factory automation. For proper operation, devices such as fluid actuators, valves, and fittings all had to be leak free. Prior to the O-ring, containment of pressurized fluids in those devices relied on gaskets or the use of complex packing seals. The O-ring, the product with 1,001 uses, replaced many of those clumsy packing seal designs with a device that was more compact and cost only pennies. The O-ring soon proved itself so versatile for the war effort that the federal government purchased the patent from Christensen for $75,000 and then turned around and offered it to the public for free.

Edwin Land (1909–1991) might be a more familiar name to some readers. His total of 500-plus patents is second only to Thomas Edison. As a Harvard graduate student in 1934, Land began his research into polarized light filters. From that work, he formed the Polaroid Corporation in 1937, making sunglasses, and later, motion picture film. Land built Polaroid into a premier research company, with the man himself deeply involved in many of the research projects. In 1948, the company released an expensive specialty camera with a 60-second self-developing film. The firm underestimated its potential market as the first production run sold out on the first day that its Model 95 camera went on sale.

## Corporate Research Centers

The earlier chapter on penicillin provided a glimpse into the wide range of laboratory resources necessary to launch a sophisticated product from its initial discovery through to its commercialization. The wonder drug began as a British effort, first at St. Mary's Hospital and then at Oxford University, the latter bolstered by British public funding and private American grants. When the program shifted to the United States in 1941, the collaboration included British scientists, foundation funding, a US Department of Agriculture lab, military funding, and finally *big-chemical* laboratories.

As industry after industry modernized, their teams of specialists displaced the individual inventor. Across the industrial spectrum, capitalists transformed their engineering approach from trial-by-error experimentation and black arts to practical science. The romanticism of genius and toil, the struggle against mountainous barriers by the lone inventor still existed—the pioneers were no longer the forefront of mankind's inventive genius.

The most well-known research programs came through World War II defense projects. First, the military-industrial complex became a showcase for the technology developed during the 1930s, and later, brandished its own newfound capacity to push innovation.

The United States military spent the largest percentage of its research funding on RADAR, an acronym for RAdio Detection And Ranging systems. Early radar detected radio anomalies using headphones. Those simpler systems required one level of technology. But presenting an object as a blip on a display screen and discerning whether it was from friend or foe, and then being able to identify the type of aircraft or ship, demanded another level of technical knowhow.

Of course, leading the way in high technology weaponry was the development of the atomic bomb. No single program of the era, and outside of the American and Soviet space programs probably ever, can match the scale of the super-secret spectacular collaboration named the Manhattan Project. The broad scope and tremendous leap forward it created in multiple sciences makes the Manhattan Project legendary. Nearly 4,000 scientists from university and federal labs collaborated in top secret. Some of those teams that turned theory into reality worked at universities such as the University of Chicago, the University of California, and the Massachusetts Institute of Technology. To finish the project, the federal government built premier laboratories in Los Alamos, New Mexico and Livermore, California. In Oak Ridge, Tennessee, the US military built an entire factory town with a population of 75,000 dedicated to the atomic bomb project.

The rapid rate of development from concept to completion, its 100% success rate, and the atomic bomb's ushering in of the nuclear age, stands as a testimony to the enormity of the operation. Yet, despite the lopsided publicity of wartime programs, private sector peacetime labs provided the majority of innovation at the time.

Pressure to either play catch-up or stay ahead of the competition drove companies to innovate. By 1939, just prior to the war machine effort, research labs employed 30,000 scientists and engineers. If a company could not afford their own research, they could reach out to one of 200 college university laboratories or 250 commercial labs. However, the big companies had their own well-staffed, well-equipped development centers.

The larger research laboratories had staffs in excess of 100 technical people representing multiple disciplines including scientists, engineers, metallurgists, and skilled technicians. The total number of employees in R&D supporting some of the largest corporations might exceed 1,000. E.I. Du Pont de Nemours operated 23 labs. In automotive, GM, Ford, and Chrysler all had impressive research centers, as did electrical industry leaders General Electric, Westinghouse, and RCA. Other notable companies included Alcoa, Dow Chemical, Eastman Kodak, Gulf, and Monsanto. Combined, their work produced automobiles, televisions, hi-tech materials, appliances, diesel engines, fertilizers, medicines, and hundreds of other products. As testimony to the benefits of research, most of those firms are still in operation today. Yet, one lab surpassed all others in size, historical impact, and with the Nobel Prize winners to back up that claim—Bell Laboratories.

## Bell Labs

Bell Laboratories of New York City developed into America's leading research facility during the Great Depression and held that position for decades following the recovery. Understanding how they became number one and how important their large scientific lab was to the country's many industries requires an appreciation for the complexities of a national and then global telephone network.

In 1877, Alexander Graham Bell and his business partners made the plunge from inventors to entrepreneurs when they launched their own communications venture. They named their new firm the Bell Telephone Company. Bell Telephone began operations with two major advantages for a startup business. First, their product had the potential for enormous demand, and second, their patent rights mitigated the threat of major competition for at least the following 16 years. However, there were also significant challenges to commercial viability.

Mr. Bell's original setup of two wired phones located a short distance from each other did not resemble the configuration of a practical telephone network. For the next step, every major component of a telephone system—the telephone, switchboard, transmission lines, and power source—all presented technical obstacles. Those hurdles seemed perpetual. Each time Bell overcame one of them they increased the popularity of phone calls, which generated more phone traffic, leading to greater engineering challenges.

A telephone was not just an upgraded version of a telegraph, a well-established technology by the late nineteenth century. Telegraphs were inherently robust. They thumped out the simple Morse code language of dots and dashes. In comparison to the telegraph, telephone signals had a delicate nature to them. Voice transmission began with one sensitive device used by the caller to generate a complex, undulating electrical signal from sound waves. The electrical signal then traveled through a wire to the listener's receiving device, which performed a near-opposite task, converting the faint electrical signal into sound waves. Even over shorter distances, voice quality in those early telephone systems degraded noticeably. For telephony to reach its communications potential, the telecommunication marvel would need lots of significant technological upgrades.

The first telephone systems were simple and independent. Each one consisting of two battery powered phones connected by one grounded iron wire. Instead of a handset consisting of a transmitter and receiver, the telephone had a dual-purpose apparatus for both talking and listening. Thus, early phones required the user to converse by alternating the handset from mouth to ear. There were no switches or telephone exchanges, just a dedicated line that was always on.

Bell made a few major improvements to the telephone during its early years. They were able to purchase the rights to a practical granular-carbon transmitter, developed by Thomas Edison, to replace Bell's finicky version. The new transmitter elevated the art of the phone conversation from shouting to talking. Then, the phone company replaced the single iron wire with two copper wires, which made for a stronger, longer traveling signal and allowed for the addition of a ringer to signal an incoming call. The telephone hook activated a switch to either engage or break the line. Bell engineers continuously worked to improve the phone's ergonomics. After all, if customers were spending lots of time on a telephone, they should be able to converse in some comfort.

In those formative years, Bell made the smart and innovative decision to relocate the battery power source, which had been attached to individual phones, to a central location where they were connected to a switchboard. By gaining control of its power grid, the phone company created a reliable energy source. Even during widespread power outages, telephones often remain operational. Those large, centralized battery packs solved one problem, but switchboards had other major obstacles to conquer.

Every phone call requires on-demand connecting wires, called lines. It is up to an operator or an automated system to make each custom connection. In comparison, a telephone system is vastly different from other utilities such as electricity, natural gas, or water supplies, which branch out from a central source to each customer. To address their unique arrangement and limit the number of electrical wires, Bell Telephone constructed a series of interconnecting exchanges, each with its own switchboards and worked by dedicated telephone operators. Bell also employed the use of shared lines, known as party lines which lacked privacy and were impractical for urban areas.

Bell Telephone still needed to solve the problem of thousands of phone calls traveling over just a handful of wires. In response, engineers invented an electronic technology named multiplexing. Introduced in the early 1920s, multiplexing divided phone calls into various carrier frequencies allowing several calls to travel simultaneously through the same line. The connection is similar to multiple radio transmitters sending their signals to many radios.

Over time, the telephone system grew in complexity. Then, after Bell's patents expired, competing phone companies emerged. Most of them formed in areas in which Bell did not supply service, although a few competed directly with Bell. In either case, Bell decided the best course was to let them into the Bell network and build special equipment to interface with the non-Bell firms. Bell Telephone operated under that system for decades.

The expanding network meant that calls had to travel farther. Therefore, a statewide, or the next step, an interstate phone system, faced larger inherent obstacles. The faint, high-frequency electric signal degraded over distance and continued to be a problem. Early electronic devices amplified the signal to a point, but eventually distortion rendered synthetic voice unusable. In response, Bell engineers, with the assistance of other researchers in the electronics community, steadily applied innovative solutions.

One critical and ongoing development was the addition of integrated electric coils into phone lines, called loading. Introduced in 1900, it reduced the attenuation of the signal and, as a result, doubled the distance that the practical voice signal could travel.

The larger effort to create clear phone signals involved the use of vacuum tubes as line boosters. They were similar to those used in radio technology.

The more precise the amplifiers became, the further a coherent phone message could travel. In 1876, the limits of a long-distance phone call extended from Boston to neighboring Cambridge. By 1925, customers could make long-distance calls between New York and San Francisco.

All those technological upgrades made telephones more popular. It seemed everybody needed a phone for their home, office, and business. They wanted to use it daily, and they wanted to talk to people everywhere. Bell customers also demanded that their phone calls be affordable, clear, and reliable. For Bell, the challenges of the telephone industry seemed infinite.

By 1925, the Bell Telephone Company was a subsidiary of American Telephone & Telegraph (AT&T). That year, AT&T teamed up with its longtime partner, Western Electric, a builder of telephone hardware, to form Bell Laboratories. Often referred to as Bell Labs, the companies created the group to establish a culture of scientific discovery and innovation capable of taking telephone technology to the next level. Located at 463 West Street in New York City, the center (now a museum) became the most prolific technical research center of the twentieth century. Bell Labs specialized in electronics and communications, but they contributed to other sciences as well. Author John Gertner fittingly entitled his book about the lab *The Idea Factory* (2013). Its many patents and Nobel Prize winners attest to that title. The engineering innovations developed there during the Great Depression now affect every American daily.

Bell Labs continued to make improvements on vacuum tubes. From 1925 to the late 1930s, the lab quadrupled a vacuum tube's efficiency and doubled its life. With the help of special undersea cables and improved vacuum tubes, transoceanic phone service reached from the US to Great Britain in 1927 and to Hawaii by 1931.

By the early 1930s, Bell Labs had invented a practical coaxial cable. A coaxial cable could handle 32 times the number of phone conversations compared to a pair of copper wires. The laboratory also developed complex electronic gadgets to increase bandwidth and allow for a greater number of calls. In the 1940s, Bell Labs began converting from manual to automated switchboards, which was another innovation designed to handle increased phone traffic. All that technology allowed Americans in 1946 to make 42 billion phone calls across 20 million miles of phone lines.

During that time, Bell Labs experimented with wireless transmissions. The lab established microwave transmission between New York and Boston in 1945 and the first mobile telephone service in 1946.

Prior to World War II, Bell Labs employed over 2,000 scientists, engineers, and technicians. They had expanded out of New York City, mainly into New Jersey. The world's largest research lab had amassed an impressive legacy of innovation, not only in telephony but also other areas of communications, electronics, and manufacturing.

Bell Laboratories served as the world's premier research facility for a considerable portion of the twentieth century. From 1925 to 1999, Bell Labs produced six Nobel Prize winners and received over 25,000 US patents. Its innovations played a significant role in modernizing America. Some of the most important of those contributions emerged during the Great Depression. Though Bell Labs focused on telephones and telephone networks, they introduced many important inventions for other industries. The following summary highlights a few of Bell's achievements during the 1920s, 1930s, and 1940s.

For movie fans, Bell Labs introduced the Vitaphone in 1926, a device used to create synchronous-sound motion picture systems known as talkies. Their condenser microphone and amplifier technologies magnified sound with minimal distortion, finding use in public address systems and in concert halls. In 1927, Bell Labs used their coaxial cable technology to transmit television images from Washington DC to New York City, the first long distance trial of its kind. By the early 1930s, they could both record and transmit hi-fidelity stereophonic sound. The lab's Vocoder, first demonstrated in 1937, synthesized speech. In the early 1940s, Russell Ohl (1898-1987) of Bell Labs invented the photovoltaic cell, known today as the solar cell.

Bell Labs also developed numerous instruments to assist the American military during World War II. Some of their most intense work focused on electrical-mechanical calculators. The Navy used those early computers to aim shipboard guns and anti-aircraft guns based on calculated ballistic trajectories. To safeguard top secret communications between Allied forces, engineers at the laboratory developed SIGSALY, an electronic speech scrambler.

The lab also made a major contribution to manufacturing. Western Electric mass produced a significant number of telephone and telephone system components. The high volumes of goods produced required sophisticated mass production methods, which created a number of obstacles. One of them was to implement systems that maintained high quality control standards without negatively impacting production rates, or in modern lingo—minimize non-value-added operations.

A Bell Labs physicist, Walter Shewhart (1891–1967), pioneered a solution using a control chart. Control charts use statistical sampling to analyze the quality of manufacturing data. The brilliance of the concept is that with a relatively small number of samples, inspectors can predict the quality of much larger populations. This allows mass production to continue with minimal interference.

Control charts serve two purposes. The first determines if a batch of parts meets specifications. If they pass inspection they move on to the next process or to the customer. If the batch fails inspection and is properly quarantined, the lot gets either scrapped or goes through a 100% inspection process to weed out the bad parts. The second function monitors the drift of any number of quality attributes. For example, a control chart may detect a gradual shift in part size due to tool wear. With the aid of statistical controls, a machine operator can make gradual, proactive tweaks instead of large, defensive adjustments, the latter often applied when the process might already be out of specification.

Later, automobile giants General Motors, Ford, and Toyota used Bell methods to develop their famous statistical process control (SPC) programs. Today's quality-based system, Six Sigma, used by many industries, also originated from Bell Lab's control charts. For the last several decades, due to manufacturing statistical methods virtually every mass-produced item costs less while maintaining higher quality.

Bell Lab's greatest invention came about through their quest to develop a superior solid-state alternative to the vacuum tube. Vacuum tubes launched the electronic age, but for the telephone industry the novel device had become a bottleneck within their vast network. Vacuum tubes were large, consumed too much power, and compared to other devices in the system, they had a short life span.

A breakthrough for an alternative surfaced when researchers in the United States and Germany discovered that metalloids such as germanium and silicon, in their pure form, could act as electrical semiconductors. As the name implies, they were part conductor, similar to copper, and part resistor, similar to wood. Eventually, the Bell Lab's Solid-State Physics Group, led by William Shockley (1910–1989), discovered that a device composed of a properly doped semiconductor topped with some delicate circuit wizardry could mimic a vacuum tube. They called their invention a transistor and received a patent for the device in 1947.

Transistors can serve one of two functions. As an amplifier, a transistor connected to a power source, such as a battery, can amplify or strengthen a weak electrical signal. That ability would make transistors a critical component in hearing aids and small radios. Transistors also are able to work as high-speed switches—the eventual building blocks for modern computers.

It would be several more years before the transistor gained worldwide attention. Transistors first found popularity beginning in 1954. At that time, several manufacturers developed radios using transistor technology. Philco supplied an automobile version for Chrysler. Texas Instruments and Raytheon built portable home transistor radios. However, it was a small Japanese company named Sony who in 1955 sold their first miniature transistor radio, the model TR-55, which helped launch the new industry.

In 1956, for their work on transistors, Shockley, along with team members John Bardeen (1908–1991) and Walter Brattain (1902–1987), shared the Nobel Prize in Physics. Shockley left Bell Labs in 1956 to establish his own laboratory in Palo Alto, California, with a focus on silicon semiconductor devices. Former employees at his laboratory spawned a number of their own high-tech companies, a process repeated by successive generations of aspiring entrepreneurs. By the 1980s, the area had become known as Silicon Valley.

The transistor ranks as one of man's greatest inventions. It forms the base for all of today's modern electronic devices, from computers, to smart phones, to robots. They are complex, but a transistor's manufacturing processes lends itself to high-speed automation, which drives down cost. Another benefit is its miniaturization capabilities. Early transistors were 1/50th the size of the typical vacuum tube—an impressive improvement at the time. However, 65 years later, the more powerful microprocessor chips contained over 5 billion transistors. As a result, today every American's daily routines at home, work, and in commerce depend upon a slew of electronic gadgets, each containing a vast array of transistors.

# 53

# The Post-War Housing Boom

In 1947, a northeast developer, Levitt & Sons, operated by Abraham Levitt and his sons William (1907–1994) and Alfred, began the construction of a 2,000-home tract in Nassau County on Long Island, New York. Their business plan was to build affordable suburban homes. The multi-faceted plan depended on volume (the 2,000 homes), a simple design (Cape Cod-styled homes), and the use of mass production methods. They named the new community Levittown.

To keep costs down, they built their houses on concrete slabs with unfinished attics. The latter was for expanding the living quarters as the family grew. Originally planned as rental properties, the developer soon recognized the high demand for home ownership and doubled the project plan to include 4,000 homes, and for the next phase they would be for sale instead of rentals. The next year, 1948, they offered 30-year mortgages, which enticed more people to buy as Levitt and sons again expanded the project. Subsequent phases included not only more homes, but additional recreation facilities and commercial buildings. By 1951, the number of Levittown homes exceeded 17,000, making it the largest mass-produced community in the country.

Besides the economy of concrete slabs, the Levitt's firm employed numerous other cost-saving techniques. They had precut lumber delivered to each site. The bill of material for each home was so complete that it included the correct number and size of nails. Teams of specialists followed an organized schedule moving from house to house performing their repeatable tasks. The system's larger vision, high efficiency, and detailed planning meant, that at its production peak, teams of builders erected a home every 16 minutes.

Levittown, and the many smaller versions constructed throughout the country in the late 1940s, left a varied historical legacy. Young families, particularly returning war veterans, could live in a safe suburban setting, if they did not mind bland cookie-cutter architecture.

One part of the Americana era that was not so revolutionary was segregation. The developers excluded both African Americans and Jewish Americans from the affordable suburban dream. Also of historical significance is the legacy of government-backed mortgages, especially the GI Bill.

Shelter is one of our most appreciated necessities. As the symbol of middle-class opulence, it is fitting that prolific housing booms stand as bookends on either end of the Great Depression. The mid 1920s real estate bubble represented both the boom and the bust of that decade, while the famous post-World War II housing explosion solidified what would become a long-lasting economic recovery.

During the 1920s, the combination of emerging technologies, such as electricity and indoor plumbing, lots of good jobs, and unprecedented easy credit provided incentives for Americans to seek better homes. With automobile ownership approaching one car per family, more people opted out of tenement living and moved to the suburbs and into single-family dwellings.

The Great Depression put the American Dream into a long-term holding pattern. But after nearly 20 years of an anemic construction industry, American new-home building swelled once again. Particularly notable were the large, suburban tracts populated with repeating patterns of ranch or Cape Cod styled homes. Historically, the home explosion, more than any other business sector, signaled economic recovery to a nation tired of war and anxious for a return to a strong American economy.

The downturn affected many industries, but construction, in dollars, dropped off a shocking 90% after the economy hit bottom and remained depressed throughout the decade. While industrial building finally picked up during the war, it was not until about 1947 that new housing starts matched those of the pre-panic housing boom. A variety of economic factors created strong incentives that would rejuvenate the new-construction housing industry. For one, the United States population during the 1930s to the mid-1940s increased by 40%. Then there was the strong economy, augmented by a lack of new construction covering almost 20 years, which meant rents and housing prices began to rise. However, it was the business model that made post-war housing recovery so impressive.

After the war, the new-home construction industry focused on value per dollar. Contractors took advantage of multiple technologies that were not available during the previous housing boom two decades earlier. They designed the next generation of homes to be practical and inexpensive to build, while maintaining high quality. Tract-home builders applied mass-production techniques similar to those used by America's assembly-line manufacturing sectors.

The design of the home itself became more practical and efficient. Floor plans included special-purpose rooms of just the right size. Eliminating wasted space reduced building costs. New kitchen designs fit the modern homemaker's needs. In an era of large families, many new homes came with more than one bathroom. Each bedroom had to have a generous-sized closet.

Homes became integrated by design. Improved, insulated wiring fed a much greater number of electrical outlets. Precision thermostats controlled either natural gas or oil-fueled central heating systems—coal was on its way out. A built-in network of either ducts or hydraulic pipes spread warmth evenly throughout the house. The uniformity of modern heating allowed for lower ceilings, a major cost saver. Insulated exterior walls added comfort and reduced the size of the heating system. Electric or gas water heaters piped hot water to a number of bathrooms, the kitchen, and sometimes a laundry room.

Home builders benefited from modern industrial engineering. The conversion from line shafting to automated, electric-powered machinery made way for a variety of manufacturers and mills to supply high-quality products at lower prices. Many of these upgrades occurred just prior to, or during, the Great Depression. The list of more affordable products included planed lumber, windows, doors, plumbing fixtures, and dozens of other housing goods.

Mankind's modern, all-around beast of burden, the truck, played a major role in reducing the cost of home building. Trucks served the factories by bringing raw materials in and finished goods out. At the construction sites, they moved everything from demolition debris to dirt and concrete. Smaller trucks carried tradesmen along with a new generation of electric hand tools.

Site preparation developed into another technological bonus for home builders. Trucks could deliver a new generation of construction equipment such as bulldozers, graders, rollers, and excavators. This labor-saving equipment built roads, dug foundations, and landscaped lawns. Machines built a network of basic services including pipes for potable water, sewage, and natural gas underground, along with electrical and telephone lines overhead.

A new class of engineering marvels, either not yet developed or underutilized before 1929, played a critical role toward constructing affordable, high-quality housing. Once again, technical advances during the Great Depression had delivered a family of products with huge cost savings yet of a high quality.

✧✧✧

Just prior to World War I, a few lumber companies, operating out of Oregon and Washington, developed the processes to produce pressed, laminated panels from thin layers of Douglas fir sandwiched with glue. Alternating the grain direction of adjacent laminations assured that the product was rigid and stable. They called their engineered sheets plywood. For years, product sales for the special lumber remained limited to home-interior doors. Then, in the mid-1930s, the industry initiated two major breakthroughs. The first was waterproof glue, which opened plywood to a much larger market. The second, standardization, which made the product more commercial by setting specifications for size, finish, and quality.

Increasing sales helped but plywood got valuable publicity from the United States military during World War II. Military uses for plywood included barracks, gliders, and boats. Plywood, though of a hardwood variety, became most famous for its liberal use on the Navy's 80-foot-long patrol-torpedo gunboats, designated as PT class boats. Plywood kept the PT boat light for high-speed travel, structurally rigid for tight maneuvering, while strong enough to handle lots of horsepower and firing armaments.

After the war, thanks to the housing boom, plywood found an even larger market. The military's use of plywood during the war proved a worthy public relations tool—in theory. At first armed forces endorsements were not quite enough for all users. Many reputable builders remained skeptics, apprehensive about a product whose strength and longevity depended upon glue. The rush to build homes overcame their concerns, which allowed plywood to gain a foothold in home construction.

Various grades of plywood found practical applications for both outside and inside the home. Exterior uses included sheathing for walls and roofing. Inside, builders used plywood for subfloors, countertop bases, and kitchen cabinets. Plywood's lower cost of installation compared to standard practices played a key role in making new homes affordable.

Plywood eventually earned the quality reputation it deserved in the construction industry. Sales after the war quadrupled compared to prewar figures and then quadrupled again over the next 5 years.

✧✧✧

Gypsum wallboard, also known by some of their popular trade names Drywall and Sheetrock, followed a path similar to plywood. Invented in 1916 for interior walls and ceilings, wallboard was a low-cost alternative to lath and plaster-covered walls. For the speedily built home, lath and plaster slowed the whole process down. The damp plaster also had the disadvantage of adding unwanted moisture into living spaces.

Despite its advantage of reducing labor costs by more than 90% compared to traditional construction methods, wallboard did not become popular until World War II. Its use caused somewhat of a revolt in the construction industry. After all, what respectable builder would replace a construction medium consisting of wood and plaster with a sheet made from chalk sandwiched between paper membranes?

During the housing boom, a skilled labor shortage *encouraged* contractors to switch to wallboard, by then available in today's standard 4 by 8-foot sheets similar to plywood. By the late 1940s, wallboard became the standard for interior wall finishing. It only took another few years for wallboard's decisive advantages to force the most reluctant quality builders to use the product.

Gypsum wallboard has remarkable strength and barely downgrades the practical structural and aesthetic qualities of a house compared to traditional methods. And wallboard has another major advantage. Whereas walls with wood slats had the potential to turn a house into a tinder box, gypsum is relatively fire resistant, thus outweighing any negatives that it represented shoddy construction.

✧✧✧

Since 1913, the Formica Corporation has produced high-pressure-laminated (HPL) sheets of Kraft paper impregnated with phenol-formaldehyde resin for both industrial and commercial applications. The company experienced only modest growth until World War II when output increased to meet military demands. After the war, the company had to find new markets.

In 1938, Formica had turned to the latest generation of plastics to find a more suitable top layer for their product. They discovered melamine, a thermosetting plastic, as the best choice. The new and improved, lower-cost laminate found a perfect application in post-war housing as a top for dinette furniture, as an alternative to wood or steel cabinet surfaces, and most famously, for countertops. Steel tended to rust, and wood was not as sanitary. Lots of cupboards and ample counter space were in high demand by the modern homemaker which drove Formica's newfound popularity.

Low cost and easy to install, Formica is scratch, moisture, and fire resistant. It tends not to trap bacteria, warp, or split. Thanks to an innovative transparent topcoat and creative lithography techniques also developed in the late 1930s, the laminate came in a wide variety of vibrant colors and patterns, including wood grain. Durable Formica products installed in midcentury homes typically performed well for decades, often long after they went out of style.

## 54

# Homemakers' Helpers

The technologies that separated the modern post-war home from the old-fashioned pre-war home were appliances. Some were built in, while others took advantage of a dedicated space and were convenient to electrical outlets and necessary plumbing. Comfort appliances included the central heating systems and water heaters mentioned in the previous chapter. Since air-conditioning was not practical for the average homeowner in the late 1940s, new homes were available with optional ventilation fans, which provided some relief during the summer months.

Kitchens became the premier showplace for the modern home appliance. Every new home *had* to have some modern appliances. Stoves, both ovens and cooktops, ran off either electricity or piped-in natural gas. Some kitchens came with a dishwasher tapped into the sink for both its water supply and its drain. The sink might even have a garbage disposal. The appliance industry set a uniform countertop height of 34 inches to help standardize the height of those built-ins. A kitchen needed room for the must-have refrigerator as well.

Laundry rooms also served to distinguish the present from the past. Washing machines and, later, clothes dryers eased the heavy household burdens of laundry day. The washing machine, along with the refrigerator were the signature home appliances that became modernized between 1930 and the post-war era.

### Refrigerators

By the late 1920s, technical advances made automatic refrigeration practical. New technologies included: scientific research of thermodynamic principles and chemical refrigerants, the widespread availability of electricity, plus the development of the efficient, fractional horsepower motor.

Previously, iceboxes achieved their chill from melting blocks of ice. A modern, thriving industry into the first few decades of the twentieth century, ice harvesters raced against the clock to cut blocks from frozen ponds, store them in icehouses, and then deliver them as needed by horse-drawn wagons to homes. Although icemen faded away with the popularity of the electric refrigerator, their cabinets, featuring finely crafted wood and thick layers of cork or sawdust insulation, formed the base for early refrigerators.

A window display at the Wisconsin Power and Light Company office features a General Electric refrigerator (Monitor Top) as "The Gateway to Health" under an arch in a picket fence. *Courtesy Wisconsin Historical Society*

Refrigeration companies supplied the electrical and mechanical components to the icebox cabinets. Then in 1927, General Electric (GE) revolutionized the industry with the introduction of its Monitor Top refrigerator, named after the mechanical assembly set on top of the cabinet which resembled a Civil War ironclad ship. At $300, GE's refrigerator was expensive for the day, yet relatively affordable considering its high level of technology. Designed as an appliance and not a piece of furniture, the Monitor Top had a steel enclosure instead of a wood cabinet. The hermetically sealed compressor, located over the cabinet, allowed for greater air flow to remove heat, which improved its efficiency. Over the next several years, other manufacturers followed GE's lead in refrigerator design. The introduction of the safer refrigerant, Freon, in 1930 (Chapter 41), encouraged even greater sales. Many monitor tops are still operational today.

As technology advanced and increasing economies-of-scale advantages lowered refrigerator prices, the market focus transitioned from the luxury market to a must-have necessity for the average household. In a competitive field, manufacturers had to get it right. Thus, industry leaders General Electric, along with Electrolux (Sweden), Frigidaire, and Philco challenged their sales forces to help design functional modern refrigerators that were better than their competition. One strategy was to sponsor traveling shows to display their wares and test design concepts on focus groups. Home economists and industrial engineers accompanied salesmen to gain first-hand knowledge of their potential customers' needs.

The finished product exemplifies the union of modern tastes and the latest technologies. Fewer trips to the grocery store led to the justification for spending hard-earned money on a new fridge. To give their customers what they wanted for storage, industry regularly increased the size of their standard refrigerator. Consumers wanted a multi-directional handle and a door that closed with minimal effort, making it easier to operate the door with loaded arms. Women wanted the mechanical unit mounted out of sight, and that the cabinet finish be smooth and painted a sanitary color such as white, although pastels later became popular. Edges needed to be rounded for safety. Homemakers wanted the top of the cabinet flat to serve extra duty as a utility shelf. Other consumer requests inspired specialty compartments such as a vegetable crisper, and trays for butter or eggs. While a single door was less expensive to build, refrigerator customers preferred a separate door for the freezer compartment, so manufacturers accommodated them with that option.

Other industries provided technologies that helped manufacturers accommodate the demanding consumer. Steels developed during the 1930s for automobiles served well for refrigerators, as did automotive sheet metal-forming methods and welding technologies. The latest polymers gave designers more valuable tools in which to meld the homemaker's demands with practical manufacturing processes. It took more than a decade, but by the war's end appliance manufacturers had accomplished their mission.

Testing a Westinghouse refrigerator-1937. *Courtesy Wisconsin Historical Society*

During the late 1940s, refrigeration appliances proved to have many benefits while they altered American lifestyles, mostly for the good. America's obsession with junk snacks came later. The refrigerator's ability to preserve foods became a huge conservation tool. For those without vegetable gardens or livestock, refrigeration provided a viable option to maintain access to fresh foods. Of course, the practicality of buying in bulk at the grocery became a societal game changer.

1947 International Harvester refrigerator. *Courtesy Wisconsin Historical Society*

Frozen foods matured as another byproduct of the refrigeration age. In 1914, Clarence Birdseye (1886-1956), a scientist working in Canada, discovered during an ice-fishing trip with the Inuit Indians that rapidly frozen fish tasted fresher once thawed and cooked. Encouraged by his discovery, Birdseye went on to develop and patent machinery designed for quick freezing. Motorized conveyors assisted the processes. In 1929, Birdseye merged with the Postum Cereal Company (Post) to become General Foods. The Birdseye division sold frozen fish, meat, vegetables, and fruit. The American public loved them, although it took a few more years after the war for frozen foods to reach full market potential. In the meantime, fleets had to equip their trucks' cargo bays with refrigeration units. Grocery stores needed to purchase the large, Birdseye-designed freezers.

Post-war, mechanical-electrical refrigeration significantly reduced the burden of household food management. For homemakers, refrigeration meant perhaps increased leisure time, further education, or an opportunity for part-time employment and some extra family income. Women had been increasingly finding more employment in the job market since the beginning of the century and as a remnant of the recent war. Modern appliances allowed that trend to continue.

Following the war, the appliance industry offered families not just one engineering marvel appliance, but two of them.

Saleswoman demonstrating a Maytag washing machine to a customer (1937). *Courtesy Wisconsin Historical Society*

## Washing Machines

During much of modern history, washing clothes and linens ranked as one of a homemaker's toughest jobs. It was known as blue Monday, itself a generous term because the many time-consuming tasks made it unlikely that a homemaker with a larger family could complete their laundry chores in one day. Plus, there were no vacations as the drudgery repeated itself week after week. Each load of laundry had to be presoaked, scrubbed, rinsed, wrung, and then hung to dry—all hard work. Hauling, boiling, pouring, and draining water added to the backbreaking and time-consuming chore. The effects of cold and hot water combined with harsh soaps took their toll on hands. Some women developed nasty cases of chilblains.

For the homemaker, technological relief came at a snail's pace, especially compared to the rapid progression of textile mill technology. On the manufacturing side, textile machinery launched the Industrial Revolution; particularly in the cotton mills where complex mechanical innovation took delicate cotton fibers and turned them into versatile cotton thread and yarn. Besides the machinery, the cotton mill itself, with its multiple processes and large scale, required sophisticated operations. It seemed that if mankind could master the manufacturing of mass-produced cotton, they could build anything—anything except a useful washing machine.

Before electricity, modern homemakers relied on hand-powered devices to ease the laundry workload. Those mechanical machines reduced the physical effort required to wash, but they did not save much time. With the spread of electrical grids, motorized gadgets replaced the hand crank. Electric washer attachments saved even more labor, but again, not so much time. Also, some poorly designed laundry machines, particularly wringers, were safety hazards, which could lead to severe injuries as they became dangerous traps for hands and hair. By 1930, the latest washing machines were safer, but the art of designing machines for washing garments still required additional, well-thought-out engineering.

The Great Depression era supplied a number of new technologies to enable development of the modern automatic washer. As with the refrigerator, its success depended upon the spread of electricity and the efficient, compact fractional-horsepower electric motor. Post-war housing builders added a few valuable infrastructures to support washers including convenient electrical outlets along with plumbed hot and cold water and drainage. Ideally, the home had all those utilities grouped into one central location such as a laundry room, kitchen, basement, or utility room.

Washing machine designers still faced obstacles to their quest for a labor-saving machine. For safety, they needed to maintain a robust separation between electricity and water. The best method of agitation continued to be a significant challenge. Washers required just the right amount of agitation—enough to clean properly, but not so much that it ripped the laundry apart. Powerful washers also needed to be tough enough to last for years, and of course, remain affordable.

Throughout the 1930s, manufacturers made dramatic advances in washing machine design. Then, World War II military production brought washing machine manufacturing to a halt. Appliance companies did not give up their quest to build a modern washing machine. Instead, still in pursuit of a vision, they turned their focus to development only. By the latter 1940s, a few appliance manufacturers released their new generation of washing machines. Lacking a form-follows-function look, washers took on our modern generic appliance style, which featured a fully enclosed cabinet made from smooth and shiny

The Bendix Home Laundry was a homemaker's dream. Bendix introduced the first automatic washer with full wash, rinse, and spin cycle at a Louisiana state fair in 1945. *Courtesy MOHAI Museum of Science and Industry*

enameled-coated steel panels. A door with a window allowed for a peek in on the washing process. Ergonomic controls made the machine simple to operate. Most important of all was a washer's automation feature. Load the basket and then with the touch of a button a network of electrical and mechanical gadgets set into motion a sequence of operations: soaking, agitating, rinsing, draining, and the beginnings of spin drying. A buzzer signaled a completed cycle. An industry in transition, lower-cost wringer washers would outsell automatic washers into the early 1950s, but the wringer's days were numbered.

As with the refrigerator, modern automatic washers fit the changes in lifestyle characteristic of a more leisure-oriented population, which in this case included clothing-specific activities—work, formal, leisure, etc. The introduction of the automatic washing machine also coincided with the beginning of another textile revolution, synthetic fabrics such as nylon and polyesters. Again, more clothing options equaled more laundry. Consider another possible modern trend relating to textiles and laundry—did washing machines contribute to the architectural layout of the post-war housing boom?

One of the major differences between post-war and pre-war housing was closet space. Not until after the mid-1940s did homeowners consider closets important. Before closets (*BC*), most families had minimalist wardrobes. For the storage of special linens and clothes, the well-to-do with enough room had wardrobe furniture. Others used trunks to protect their precious textiles. But for most, clothes just hung on wall pegs. If houses had closets, they were too small and too few for modern standards. People considered them a "space behind a door."

Once clothing storage became fashionable, then buying a house with a spacious closet in every bedroom and a linen locker for each bathroom upped the incentive to purchase a new home—preferably in the suburbs. Finally, 150 years after the birth of the cotton manufacturing boom, inexpensive textiles became a practical purchase, which is what Arkwright, Slater, and other textile engineering pioneers intended.

The appliance laundry industry embodied the modernization of America during the 1930s. In 1939, they sold 1.5 million washing machines, four times the volume compared to 1929 sales. Over that span, the wholesale price of a washing machine plummeted from $146 to $36. Plus, there was the tremendous added value of modern features. To the Keynesian academic, those numbers equate to near zero-growth over a ten-year span, but to the historian, the progression of the washing machine should represent an underappreciated economic revolution within the Great Depression.

# Part Eight

## Lessons Not Learned

# 55
# Recovery by Millions of Know-Hows

In 1958, Leonard E. Read (1898–1983), founder of the libertarian think tank the Foundation for Economic Education (FEE), published a brilliant essay titled "I, Pencil." Read's 2,230-word exposé describes the production of the ordinary wooden pencil, and in the process, impresses on his readers how a common object that appears to be so simple is actually quite sophisticated. The author used the story to teach two profound lessons about capitalism and, although not his intention, he educates us on how America made a dramatic recovery from the Great Depression.

Read's first goal was to explain the complexities of the simple American-made pencil. He began the lesson with an overview of the vast array of materials required to build a 1950's-era pencil, some of those materials collected from the far corners of the globe. The wood shank originated from the straight-grained trunks of cedar trees grown in northern California and Oregon. The lead, which did not contain the heavy metal, consisted of a complex recipe of graphite from Sri Lanka, refined ammonium hydroxide from Mississippi clay, candelilla wax from Mexico, paraffin wax, and animal fats. The main ingredient for the familiar yellow lacquer was castor bean oil. For the printed label, pencil makers used carbon black, famous for making tires more durable after the turn of the century. The eraser, or the *plug*, as the industry refers to the pencil's crowning glory, uses a rubber substitute called factice to improve erasing. Factice contains rapeseed oil from Indonesia, and sulfur chloride. The formula only needs vulcanized rubber as a binder. The pink color comes from cadmium sulfide. The shiny ferrule that holds the eraser, contains mainly brass, an alloy composed of copper and tin. The ferrule's etched dark rings use blackened nickel.

Read also detailed some logistics and manufacturing processes involved in pencilmaking at a time when American manufacturers produced 1.5 billion pencils annually. The transportation needed to bring the pencil's raw materials to factories presents a major logistical undertaking. A completed pencil, with all its launch points, terminals, and destinations, relies on a coordinated network of many ships, trains, trucks, and handlers. As with any manufactured product, pencil making requires both standardized and specialized machinery. A few of a pencil's manufacturing processes include sawing, baking, and molding. To keep production lines flowing at high speeds, manufacturers use an assortment of custom assembly and packaging systems.

But how do pencilmakers get the graphite inside the wooden shaft? Saws and planers fashion the wood into 7-inch long slats, each 2 inches wide and just over 1/8 of an inch thick. Special machines then cut 8 half-round grooves in each slat. Another operation places the extruded spaghetti-like graphite into those grooves. To form the shank, machines mate and glue two slats together, one with graphite and one without. Routers then milled the hexagon shape to create 8 pencils from each block.

L.E. Read, with *I, Pencil*, makes two perceptive points about the complexity of an ordinary wooden pencil. First, he emphasizes, referring by the first-person, "Simple? Yet, not a single person on the face of this earth knows how to make me [a pencil]." That assertion by Read includes a pencil company's president and its most experienced engineer. Each worker in the industry has their specialty and at best, only a cursory knowledge of other processes and no knowledge of most. Nobody has a grasp on all the hundreds of little know-hows, a catchy term Read used to describe industrial talents required to manufacture a pencil.

Read's second point is even more incredible. The multifaceted making of a pencil does not require a mastermind. Instead, as Read explains, "We find the Invisible Hand at work." Thus, the massive coordination effort does not depend on guidance from any government. Pencil manufacturing does not need federal trade agreements, oversight committees to promote fair competition, or research grants. This insight is contrary to many people's natural instincts, yet basic to the concept of robust economic growth and liberty.

Read uses his pencil story as a stepping stone to encourage the reader to appreciate other everyday products with complexities significantly greater than a pencil's, such as a typewriter, a grain combine, an automobile, a milling machine, a jet airliner, or the many thousands of other products manufactured at the time.

Milton Friedman retold a version of *I, Pencil* in both the PBS television series (1979) and in his book titled *Free to Choose*. In a television interview, Dr. Friedman said the following about Leonard Read's insightful article.

> "I, Pencil", has become a classic, and deservedly so. I know of no other piece of literature that so succinctly, persuasively, and effectively illustrates the meaning of both Adam Smith's invisible hand—the possibility of cooperation without coercion—and Friedrich Hayek's emphasis on the importance

of dispersed knowledge and the role of the price system in communicating information that will make the individuals do the desirable things without anyone having to tell them what to do.

The pencil story, with Mr. Read's clever term "know-hows," explains a couple more valuable lessons in economics, including some insights into how the American economy emerged from World War II considerably stronger than where it stood when the stock market crashed in October 1929.

As a tool to understanding economic growth, the concept of Leonard Read's know-hows serve as "opportunities" within the capitalist system. All the phases of manufacturing, along with design, logistics, etc., contain many know-hows that encompass overwhelming innovation potential. Even for the pencil, Read covered but a small fraction of a pencil's processes to make his case. Consumers during the 1930s had access to tens-of-thousands of products, each one composed of anywhere from several to thousands of components. Those components required between a few to dozens of processes. All totaled, many millions of know-hows powered the American economy. Consider that each day, some workers of an unknown number, equipped with various skills, put in that extra effort to advance a traditional know-how or create a new one. For the 1930s, the private sector employed an average of 45 million workers, a force plenty large enough to form a labor pool of innovators. Most of those improvements were small: perhaps a clever way to shave a half of a second off the time to assemble a toaster. Others were more revolutionary such as the developments that led to the practical diesel engine. All that continuous improvement lies at the heart of the American capitalist system—an economy that constantly upgrades as it moves forward. Even better, the escalation of know-hows helped make the American economy modern.

For society to be "modern" was not the introduction of humankind's many wondrous inventions (scientific discovery), it was the consumption of advanced versions of those technological marvels by the masses. Volume is the hallmark of a modern economy, which gives innovation leverage. Remember, Thomas Edison's breakthrough light bulb, patented in 1878, added no wealth to the nation's economy. Twenty years later, light bulb advancements still had a minimal impact on the average American's economic wellbeing. After 1910, under the control of GE, upgrades such as an improved glow, longer life, a lower cost, and most important, the construction of light bulb factories, leveraged innovation over millions of light bulbs.

By the way, for the 1930s, chemists tweaked a pencil's lead recipe to make it stronger and less prone to breaking. Imagine the added convenience of reducing the number of times the point breaks over the life of a pencil. To appreciate the economic impact, apply that benefit towards some billion-odd pencils annually—just one small example of the extraordinary impact some simple know-hows can have on the nation's quality of life.

In another example, the opportunity to invent the automobile had already passed by the turn of the twentieth century, but the opportunities for improvements were infinite During the Great Depression hundreds of innovations added tremendous value to cars such as improved safety, convenience, reliability, performance, comfort, and drivability. From the 1930s into the mid-1940s, the list of reinventions also included washing machines, refrigerators, toys, packaging, building materials, and countless numbers of other products made in high volumes.

How did academia address the 1930s industrial revolution? Towards the end of the Great Depression, universities added the field of macroeconomics to their economic curriculum. Macro-economists tasked themselves with the study and analysis of economic growth. The timing was perfect since the era served as an ideal economic laboratory. Compared to the latter-twentieth and twenty-first centuries, the dollar maintained long-term stability, the federal government's influence was much smaller, and while industry continued to modernize, it still had a form-follows-function element to it compared to the microelectronics influence of future decades. For academia, it was a chance to embrace the elegance of capitalism, delve into its tremendous power, and to appreciate modernization's egalitarian elements.

Instead, despite the latest economic transformation unfolding before them, the macro economist ignored the wonders of American exceptionalism. Perhaps overwhelmed by industrialization's range and depth, or just unimpressed, they turned their backs on real growth and retreated to their ivory towers to over-analyze their newfound superstar Keynesian economics, which monitors money excesses and not real wealth. Most historians blindly followed their hapless colleagues on an 80-plus-year journey of missed opportunities.

# 56

# Questionable Lessons

The American economy of the 1930s morphed from what should have been a routine economic recession into the infamous Great Depression. In the panic's aftermath, academia then spent the next 90 years trying to determine what went wrong. The quest continues to haunt some historians. From all that research, did academia learn some valuable lessons that could help avoid future economic catastrophes? The short answer: Not really.

### Trade

When President Hoover signed the protectionist Smoot-Hawley Tariff Act in May 1930, an already sluggish economy plummeted. As America's global trading partners retaliated with harsh import duties targeting the United States, the economy hit bottom. By the time of the recovery, seventeen-odd years later, historic free trade agreements had replaced punitive tariffs. The destructive nature of Smoot-Hawley's over-zealous, punitive import duties sent a strong warning to politicians, bureaucrats, pundits, and academics from both sides of the political aisle regarding the dangers of using high tariffs to bail out a struggling economy. Due to its dreadful reputation, the desire to avoid a Smoot-Hawley repeat might last for centuries.

Today, academia's aversion to high tariffs as a recovery tool serves more as a reminder than as a true lesson learned. Remember, the New York Times May 5, 1930 editorial, authored by 1,028 leading economists, pleaded for either Congress or the President to kill the protectionist bill. Plus, there were those industrialists, financiers, and confidants who visited the oval office to dissuade Hoover from signing the disastrous trade initiative.

### Safety Net Programs

One noticeable civic difference between the beginning and the end of the Great Depression were the many federal safety-net programs enacted to shore up the income security of the nation's most vulnerable citizens. Those programs addressed economic slumps and provided aid for the poorest individuals and families regardless of the economy's condition.

Many of these programs developed from initiatives passed under the

Roosevelt administration are still in use today. Unemployment compensation offers financial relief to those who lost their jobs and have difficulty finding another. The benefits, even with extensions, work only for the shorter panics as they fail to cover longer-term recessions. Social Security, a New Deal staple, provides a reasonable income for senior citizens and those who become disabled before retirement. FDR used his three-legged stool metaphor to promote his pragmatic view of Social Security. The President noted that the three legs represented pensions, savings, and Social Security. For a more comprehensive layer of protection, federal politicians created numerous government programs for the poor known as welfare. To assist the number of people in desperate need of help, Congress increased funds for food stamps, housing subsidies, and subsistence income for the poor, all geared toward families with children. The belief was that without the federal umbrella, some might fall through the cracks if they had to depend upon traditional public or private aid. Despite the large-scale humanitarian effort, these federal initiatives promoting income security deserve, at best, only partial credit as a Great Depression lesson learned.

For one, was the national welfare system in 1930 broken? Prior to the New Deal the United States, had an extensive aid network funded jointly by private charities and local communities. American churches played a significant role in that charity network. Up to a point, they worked well. For example, although plenty of people went hungry, aside from the occasional isolated tragedy which can occur during any decade, there was no starvation due to the Great Depression. True, the social burdens of the Great Depression overwhelmed the system, but the reckless behavior of a more intrusive federal government caused the bulk of the economic devastation.

Progressives were able to shift the blame for the infamous recession to a lack of responsible government, whose antidote toward recovery was to create a larger central bureaucracy. Then they took the primary responsibility for the nation's social welfare. Deflecting the culpability of the Great Depression on the private sector was a difficult caper to pull off, but they did it.

One major factor that made the United States the richest and most advanced nation in the world was the elevated power granted to its numerous states. The limited role of the federal government helped entitlement programs avoid much of the waste associated with central bureaucracies and provided the system with competition and diversity. After the New Deal era, America began to lose that advantage as various entitlement and safety-net programs switched to less efficient one-size-fits-all solutions. For example, the enormous size and scope of welfare competed against private charities leading to the demise of countless well-run charitable organizations, a trend known as crowd-out.

On top of this, Congress, along with certain presidents, eventually crossed the line from maintaining a passionate civil responsibility designed to help the disadvantaged to creating a dependent lower class in exchange for votes.

## Taxation

Tax policy is another issue where twenty-first century politicians appear to have learned a lesson from the Great Depression. Personal, corporate, and excise taxes swelled during both President Hoover's and President Roosevelt's administrations. Today, the trend favored by politicians from both sides of the aisle when the country is in the grips of a recession is to cut taxes. Ronald Reagan (1911–2004) cut taxes in response to the 1981 recession. The George W. Bush tax cuts intended to stimulate the economy during the 2008 Great Recession, and extended by President Barack Obama, are more recent examples. But were those real tax cuts?

Today's custom of perennial deficit spending has purged the need for emergency rainy-day funds. As a result, the responsibility to compensate for lost tax revenues calls on the US Treasury to be creative. In fact, modern recessionary fiscal policies stress federal balance sheets already in disarray. Besides tax cuts, the federal government loses tax revenues due to the recession, and they must pay for pump-priming spending, another lesson-not-learned given that ten years of stimulus failed to rejuvenate the 1930s economy. To pay for the mounting shortcomings, treasury accountants take advantage of the nation's liberal fiat monetary policy and create new money. This process inflates both the national income and the amount of federal spending equally. Consequently, federal spending accounts for a higher percentage of national income, and therefore, the printed money, in the form of inflation, becomes a tax increase.

Perhaps, tax cuts are good for the economic soul of the nation. If so, a modern recession-based taxation policy counts as another example of partial credit for lessons learned during the Great Depression.

## Intro to Lessons Not Learned

A longer record in what should have been a productive scholastic discovery process is the profound "lessons not learned" list. The lack of a better understanding of the Great Depression comes in part from those embracing left-leaning political agendas, dominant in universities since the end of the war. After all, "History is written by the victors." The bulk of bad scholarship originates from a field that should have best grasped the economic intricacies of those times. Economists, just prior to the start of the recovery, devised a scheme in which they counted generous amounts of new fiat money, a popular trend now and then, as a substitute for real wealth. The ruse helped to mislead historians and policy makers about the nature of economic growth and, as a result, has plagued the study of the Great Depression ever since.

# 57

# Measuring Grossly Distorted Productivity

For three weeks during July 1944, political and financial leaders representing the 44 Allied nations met for the United Nations Monetary and Financial Conference at the Mount Washington Hotel in Bretton Woods, New Hampshire. Their primary goals were to establish currency rules along with free trade agreements and other initiatives designed to foster global economic prosperity. Known today as the Bretton Woods Conference the historic meetings led to the formation of the World Bank Group (WBG) and the International Monetary Fund (IMF). To measure the output of their lofty goals (after all, they did not want their good deeds to go unnoticed.) Bretton Woods delegates adopted an economic indicator known as the gross domestic product (GDP). According to the IMF,

> The GDP measures the monetary value of final goods and services – that are bought by the final user – produced in a country in a given period of time (say a quarter or a year).

A popular snapshot of economic wellbeing known as the *equation of exchange* defines the GDP in mathematical terms. The expression relates a nation's money supply and monetary velocity to the quantity of goods and services multiplied by the average price of those goods and services. Packaged in a relatively simple formula, the elegant monetary mathematical identity states that:

$MV = PQ$
Where:
$M$ = money supply
$V$ = monetary velocity (circulation rate)
$P$ = consumer price index (CPI)—average price of all final goods and services
$Q$ = quantity of all final goods and services
$PQ$ = gross domestic product (GDP)

One cannot overstate the absurdity of the GDP as a growth metric. Figures 57.1 and 57.2 (described in more detail below) illustrate the equation of exchange in graphical form for two eras—the Great Depression and into the twenty-first century. MV is the amount of money circulating about an economy. PQ is the aggregate of all the stuff that money buys. Nowhere in the equation is a hint of its real worth. Instead what those graphs show is inflation masquerading as growth.

Figure 57.1 tracks the GDP, along with its two sides of the equation of exchange through the Great Depression years. M, shown as M2, includes both money and *near money* such as checking accounts and mutual funds. Velocity (or money circulation rate) is the number of times per year a dollar purchases a final good or service. For simplicity, the value of velocity used in the charts is 2.0—its approximate long-term average. The chart matches the mainstream version of the economic panic. The economy slumps after 1929, followed by a modest recovery from the New Deal, then a meteoric rise attributed to World War II spending, and finally the post-war recovery.

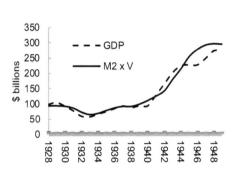

Figure 57.1 Money Supply (M2) x Monetary Velocity (V) Vs. Gross Domestic Product (GDP), 1928-1950

By the 1950s, the economy settled down to its current perceived ever-increasing growth rate—Figure 57.2 (since 2007, the Feds aggressive devaluation of the dollar has brought M2 velocity down to 1.5). Based on the shape of the two graphs, its mathematical nature, a legitimacy based on reams of data, and its Keynesian elements—the concept of a GDP caught on.

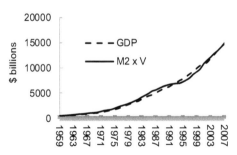

Figure 57.2 Money Supply (M2) x Monetary Velocity (V) Vs. Gross Domestic Product (GDP), 1959-2007

This economic indicator had seen some use in the US beginning in 1937, then after Bretton Woods, the GDP became an essential economic tool (Older sources often use the gross national product, GNP—slightly different criteria yet with the same faults). Today, the US Department of Commerce's Bureau of Economic Analysis (BEA) releases quarterly GDP statistics along with updated adjustments to account for inflation. The perceived accuracy of the BEA's analysis of the GDP is implied by the published metric's 0.1% resolution. US federal agencies, aided by the Federal Reserve System, target about 3.0% as an ideal GDP growth rate. According to conventional wisdom, above a 4.0% GDP increase indicates an overheated economy and an impending recession. Below 2.0% indicates a recession in progress.

The gross domestic product has become a staple of every news organization's business reports. At a research level, scholars apply advanced mathematics to the concept in order to help guide policy makers toward unlocking the secrets of the *ideal* growth rate for long-term economic expansion. A few of those gallant efforts have won researchers Nobel Prizes. Furthermore, GDP principles are core to macroeconomic curriculums.

For all its hype and significance, the GDP's most notable attribute is that it has **zero** ability to fulfill its primary design objective—gauge economic growth. Otherwise, the GDP does have the crude ability to compare the ratio of government spending to private-sector spending. Yet, somehow the GDP continues to be the authoritative metric for measuring economic growth. Also, the GDP is also valued as a troubleshooting tool for analyzing the nation's economic misery during the Great Depression and its stunning recovery after the war.

How a metric with such universal appeal could be so worthless seems incredible, but there is ample evidence to back that claim. In fact, the uniqueness of the period covering the 1920s through the 1940s serves as an ideal laboratory to expose the GDP's faults. For one, the era had periods of both low and high inflation. The extensive use of mass production came of age. Finally, there was the technological proliferation that transformed many consumer, commercial, and industrial goods.

One widespread problem that has corrupted the *real* worth of the GDP is mass production. Any industry's shift from shop to factory increases volumes while lowering prices (PQ from the monetary of exchange identity); therefore, as a product of the two, industrialization tends to create real growth not reflected in the GDP. However, that trend has gone unnoticed by academia. Then probably more by serendipity than design, just as mass production began to flourish across a slew of industries, new-age public strategy makers turned to the Keynesian doctrine as the Englishman's easy-money strategy became the rage. The influx of new dollars more than offset industry's cost reductions and allowed economists to model an economic system that had a natural tendency to expand, but at a rate that had no relationship to reality.

As noted, the era spanning the first four decades of the twentieth century serves well to underscore the absurdity of the *lab-grown* metric. Leading up to and into the early stages of the Great Depression, the volume of goods increased while their prices decreased. Because there was low inflation, the GDP remained underwhelming—masking the output of a revolutionary vigorous economy. Similar trends caused the 1930s economy to appear more stagnant than it actually was. A few examples demonstrate the misunderstanding (where both volumes and incomes have been published).

Table 57.1 compares the 1929 GDP of household washing machines to those of 1939. Despite a whopping four-fold increase in the annual number of household washing machines produced, the more relevant growth metric which were dramatic cost reductions, decreased their portion of the GDP by 24%. Real growth would be closer to an increase of 307% annually. In the case of a washing machine, a valuable Great Depression product beneficial to the working family, gets penalized per the GDP for being more affordable.

**Table 57.1** Household Washing Machine Sales

| Year | Volume | Sales $ | % Δ GDP | % Δ Volume |
|---|---|---|---|---|
| 1929 | 370,000 | 71,000,000 | | |
| 1939 | 1,505,000 | 54,000,000 | -24 | 307 |
| annual averages → | | | -2.4 | 30.7 |

Not as dramatic a change as washing machines, automobile sales of the famed Model T from 1908 through 1916 (Table 57.2). In all but the last data bin, automobile sales outpaced the theoretical GDP.

The Model T table also busts an automation myth. According to standard history, the big changes in automobile production occurred after the introduction of the moving assembly line in 1913. The numbers in the Model T table challenge that claim.

**Table 57.2** Model T Sales

| Year | No. Sold | $- thousands | % Δ GDP | % Δ Volume |
|---|---|---|---|---|
| 1908 | 10,660 | 10,127 | | |
| 1909 | 19,051 | 14,860 | 47 | 79 |
| 1910 | 34,070 | 23,508 | 58 | 79 |
| 1911 | 76,150 | 45,690 | 94 | 124 |
| 1912 | 181,951 | 100,073 | 119 | 139 |
| 1913 | 264,972 | 129,836 | 30 | 46 |
| 1914 | 283,161 | 124,591 | -4 | 7 |
| 1915 | 534,108 | 192,279 | 54 | 89 |
| 1916 | 785,433 | 282,756 | 47 | 47 |
| annual averages → | | | 56 | 76 |

Table 57.3 tracks wheat production for the years leading up to the Great Depression. Even though a subcomponent of the GDP, the fall in wheat prices again challenges the validity of the economic metric. The negative GDP corresponds with claims that a collapsed agricultural sector helped cause the Great Depression, but the

**Table 57.3** Wheat

| Year | Bushels x 1,000 | Sales $ millions | % Δ GDP | % Δ Volume |
|---|---|---|---|---|
| 1925 | 669 | 961 | | |
| 1930 | 887 | 595 | -38 | 32.6 |
| annual averages → | | | -7.6 | 6.5 |

healthy increase in volume suggests the opposite—a vigorous American capitalist economy. Other farm products showed similar results.

Per Table 57.4, bituminous coal prices plummeted throughout the 1920s. While prices plunged, volumes just tapered off. Those figures were good for consumers, but lower prices signaled bad news for certain workers, which might help explain why it was a turbulent time for coal mine labor unions.

Table 57.4 Bituminous Coal

| Year | Volume-tons | Sales $ millions | % Δ GDP | % Δ Volume |
|---|---|---|---|---|
| 1920 | 569 | 2,130 | | |
| 1925 | 520 | 1,060 | -50 | -8.6 |
| 1930 | 468 | 795 | -25 | -10.0 |
| annual averages → | | | -7.52 | -1.86 |

The problem is that economists are not gauging real wealth but instead are counting money flowing through the economy. The tight association between the PQ (GDP) and MV makes sense, but in true Keynesian theory :

- price ($P$) and velocity ($V$) should remain constant
- quantity of goods and services ($Q$) increases
- money supply ($M$) increases to both encourage growth and maintain financial stability

In practice, per the previous tables:

- $Q$ expands
- $P$ plummets
- $M$ inflates the GDP as it also allows for deficit spending

Economists could conserve significant resources calculating the GDP by counting money instead of tallying market data.

If a volume-versus-price relationship was the sole issue associated with the GDP then economists might be able to sort that out, but there are much larger problems.

Technology-motivated price cuts were just one factor distorting the GDP—upgrades proved to be another. For example, the GDP has no ability to account for added value. Attributes such as greater utility, improved durability, added safety, and superior aesthetics carry no weight with the popular economic indicator. As many of the products featured in earlier chapters chronicle, powerful visuals from the form-follows-function era demonstrates America's dramatic transformation from old-fashioned to modern. The washing machines featured in Chapter 54, reinforce that assertion. Each generation of appliances fall under the category of washing machines because their function is to wash clothes. Otherwise, they share little in common. The earlier models proved to be labor saving, but they were not time savers. By 1947, a homemaker with a state-of-the-art washer had the advantages of both. Load the basket with dirty laundry, push a button, and wait. Nowhere does the GDP account for those conveniences.

The photos of four Ford automobiles featured below (Table 57.) further discredits the GDP as a measurement of economic wealth or growth. The 1930 Model A represented the latest automotive technology at the start of the Great Depression. Over the next two decades—which covers the ongoing recession, World War II, and the economic recovery—there appears to be no relation between automobile price increases and value-added improvements in styling, safety, performance, drivability, utility, and durability. Those changes represented thousands of profound advancements in mechanisms, electric gadgets, industrial processes, manufacturing automation, and material sciences.

**Table 57.5** Progression of the Automobile

| 1930 Ford Model A | 1939 Ford | 1946 Ford | 1949 Ford |
| MSRP $495 | MSRP $681 | MSRP $1138 | MSRP $1425 |

The price of a basic Ford rose slightly between 1930 and 1939, but over the same period there was a dramatic rise in engineering (see Chapter 33). Conversely, the differences between the 1939 and 1946 models, despite a healthy price increase, are mostly cosmetic. Then back to a bargain, as the revamped 1949 Ford (debuted June 1948) featured numerous, significant upgrades (see Chapter 51) which its modest price increases fail to capture.

To the economists, an automobile is an automobile. They are incapable, through the GDP, of distinguishing the difference in value between a turn of the century motorized buggy and the sleek 1949 Ford. The only thing that matters is price and volume. As noted, if prices half and volumes double, a sign of real economic growth, then the GDP remains flat.

New products, whether they may or may not have an older equivalent, reveal yet another shortcoming of the gross domestic product. In a not-so-hypothetical side-by-side comparison, a billion-dollar snake-oil industry has the exact same value as its *replacement*, a billion-dollar penicillin industry. Within

GDP land, a tonic—which might work as a placebo at best—has as much worth as a drug that saves thousands of lives per year. Other new releases during the Great Depression and the recovery included: bulldozers, plastic goods, forklift trucks, Skilsaws, and televisions. Economists cannot quantify the economic value of those products, or the worth of many others, by calculating the aggregate of their prices and volumes.

Another problem with the GDP is that technologies are not always beneficial for economic wellbeing. Many products and processes can be dangerous to human health, as they might cause injury, foul the air, pollute waterways, poison our food, or deplete the earth's ozone layer. In a couple of modern examples, what are the ultimate benefits and costs of opioids or violent video games? Money spent on rescue, cleanup, and rebuilding due to natural disasters such as hurricanes, earthquakes, floods, and tornadoes present similar problems with the GDP, as does counterproductive or frivolous government spending.

A dependency on the GDP as a tool for measuring economic growth ruined Great Depression histography. For the 1920s, a strict gold standard led to an understated GDP. The flat GDP then diverted historians' attention away from industry's latest technological revolution including the proliferation of an electrical power grid, the internal combustion engine, electronics, and the motorized factory. In response to the decade's *low* monetary growth rate, analysts began to overplay various financial issues as the primary causes of the Great Depression.

The 1930s played out in a similar manner. A lackluster GDP creeped up from 1933 until the end of the decade, which was—as noted throughout this book—a response to a combination of slight monetary expansion and some increases in consumer confidence. That GDP trend turned out to be enough evidence to judge the decade as having mediocre growth. With a fixation on GDP instead of real progress, academics ignored the many pockets of industrial might and instead turned to an unhealthy focus on the New Deal as the panic's savior.

The GDP soared from 1941 through 1945, yet the severe shortages and the ravages of war contradict the standard metric as a sensible definition of legitimate economic growth. Nevertheless, handicapped by academia's blind spot, scholars turned to the GDP to elevate the World War II military-industrial machine as the de facto source that ended the Great Depression.

With the economy back on track, plenty of evidence indicates that real industrial growth far outpaced GDP. Nevertheless, because of standard economic practices, historians tell us that clearing walking trails, the construction of hydro-electric plants in rural areas, the military buildup of tanks, planes, and ships combined to produce diesel-electric locomotives, affordable, high-quality refrigerators, nylon stockings, and hundreds of other high-profile products that modernized the economy.

The radical transformation of American industry over the last century should have exposed the GDP's weaknesses. Instead, not only has the ongoing ruse corrupted Great Depression historical analysis, the flawed measurement continues to plague a full appreciation for today's real national economic prosperity.

# 58
# A Useless Appendage

Today, much of the Federal Reserve System's monetary policy is a response to the financial struggles endured during the Great Depression. Readers might recognize some of these from the Fed's official mission statement:

> The Federal Reserve System is the central bank of the United States. It performs five general functions to promote the effective operation of the US economy and, more generally, the public interest. The Federal Reserve:
>
> - conducts the nation's monetary policy **to promote maximum employment, stable prices, and moderate long-term interest rates** in the US economy;
> - promotes the stability of the financial system and seeks to minimize and contain systemic risks through active monitoring and engagement in the US and abroad;
> - promotes the safety and soundness of individual financial institutions and monitors their impact on the financial system as a whole;
> - fosters payment and settlement system safety and efficiency through services to the banking industry and the US government that facilitate US-dollar transactions and payments; and
> - promotes consumer protection and community development through consumer-focused supervision and examination, research and analysis of emerging consumer issues and trends, community economic development activities, and the administration of consumer laws and regulations."

Outside of shoring up the nation's private banks, one of the Fed's primary goals is to smooth out the business cycle, and, as a byproduct, achieve steady growth (see underlined section above). The US Treasury shares a similar goal—as underlined in their mission statement:

> Treasury promotes economic growth through policies **to support job creation, investment, and economic stability**.

Treasury also oversees the production of coins and currency, the disbursement of payments to the public, revenue collection, and the funds to run the federal government.

The modern economist believes that a powerful central bank has the tools to promote both short-term and long-term economic wellness by smoothing the business cycle. The practice appears routine. Tighten the money supply during a bubble and then prime the pump at the onset of the slump. In truth, there are numerous, fundamental flaws with the Fed's scheme to manage growth with dollars. First, to tighten the money supply in order to prevent an impending bubble requires an excess of funds. In effect, the treasury needs to manage a surplus to have something to cut back on—a policy of perpetual inflation. This was a ploy the central bank had to work up to. In the 1920s, America had a strong dollar, thus the federal government would need to reverse their decades-old tight-monetary policy for an easy-money policy, which is what the progressives sought. To fill its coffers, the Fed partnered with the Treasury to increase the money supply with fiat cash. However, tweaking an energetic yet fickle economy is not that easy. If the Fed subsidizes its currency by a small amount, their efforts lack the oomph needed to stymie runaway consumer confidence—the force behind the bubble. If the Fed adopts a liberal dose of Keynesian economics and prints lots of dollars to increase their leverage, then they might set off alarms that could expose their treachery. Then comes the tricky part. Is a case of heightened economic activity due to a surge in consumer confidence or the result of too much easy money?

The next problem is the ability of the Fed to react to a bubble. Realistically there is a small window, if any, for analysts to recognize the formation of a bubble, act on it, then implement a damage-control strategy before the crash. The Fed tools might not be as powerful as advertised. Despite the Fed's best efforts, higher interest rates (the preferred tactic) might carry over into the slump. To recap. Raising interest rates to calm an overexuberant economy when it arises requires that the federal government constantly weaken the dollar.

Even stimulus management has its problems. Unlike the sudden onset of a crash, the recovery phase is a long and slow process. The timing allows for lots of time to do the right thing, or the wrong thing. The larger problems are again market forces and the need for significant spending which places the burden on the Fed to print money as politicians seek to enact stimulus without taxation.

Business cycle bubbles coexist with a nationwide shopping spree. The crash that follows results from austerity brought on by those increased expenditures—the reality of overspending. For example, a family caught up in an exuberant economy might purchase an automobile they cannot afford. Stimulus does not change the fact that they do not need another car or that they now have expensive car payments due to their buying indiscretion. As the buying frenzy flows across the economy, the Fed's stimulus will most likely fail to produce a real recovery.

The Federal Reserve System uses smooth-the-business-cycle operations for another scam—the illusion of near-continuous growth. They combine the faulty GDP indicator with liberal fiat funds to validate the success of their enterprise. Therefore, through deception, the Fed usually achieves their vision of an ideal-growth curve—a political self-fulfilling prophecy.

The Fed's ruse has ironclad support. To politicians, devaluing the dollar is a lazy version of ancient Rome's bread and circuses policies. Instead of goods and services, the central bank doctors the federal government's ledger. Academia embraces the concept because they designed it. Unfortunately, academia, as both the architects and the censors of the ploy, back a system that lacks legitimate oversight.

It is the market's function to flatten the business cycle. The system is not perfect, but it works within the boundaries of fickle human nature. Basic financial instincts advise consumers and investors to sell stocks, goods, and services when prices are high and buy them when they are low. Over time, experience hardens the process. For continuous economic expansion, innovation and commerce create growth while fiat dollars do not.

The US Treasury has the responsibility to maintain a strong dollar, but that is a conservative ideal. For the left, whose principals require a weakened dollar, there exists a powerful thought—the fear of another Great Depression. To that aim, and sadly, as a puppet of the political left the central bank's primary function—perpetually inflate the dollar.

Since the time of the Great Depression, America's financial leaders have depended on the Keynesian model to drive national fiscal policy. That policy strives to balance excessive deficit spending against acceptable inflation. In reality, America's fiscal leaders have succeeded in diminishing the tremendous power of the nation's industrial sectors

The scheme's absurdity becomes apparent during recessions. Consider the effects of a deficit-spending economy following an economic downturn. An economy, as it will do periodically, surges and creates an economic bubble, then the bubble bursts, which results in a recession. The recommended medicine to motivate the downturn is to prime the pump with federal spending. In response, the treasury generates large amounts of money to fund those stimulus packages. With enough outlays, no matter how frivolous, the GDP rises, the stock market rises, and even before unemployment returns to normalcy, the economy is technically recovered, although their mission is accomplished by inflation only.

In her lecture, "Lessons Learned from the Great Depression," Christina Romer (b. 1958), a former Chairperson of the Council of Economic Advisers under President Obama and a leading authority on the Great Depression, exemplifies this easy-money stance with her five fiscal recommendations for economic recovery in the case of a recession.

- A small fiscal expansion has only small effects
- Monetary expansion can help to heal an economy even when interest rates are near zero
- Beware of cutting back on stimulus too soon
- Financial recovery and real recovery go together
- Worldwide expansionary policy shares the burdens and the benefits of recovery

Professor Romer's advice echoes conventional wisdom's policy regarding money—print early and print often. It is difficult to distinguish whether their stand characterizes a critical mass of naive economists or a stealthy policy executed by brilliant progressives. Either way, most intriguing is that the easy-money strategy is self-fulfilling.

Again, regardless of how destructive the government's response is to the economy, with enough monetary fabrication, fiscal leaders can declare their methods successful. Thus, a recovering economy can be both mired in bureaucratic waste, yet statistically, via increases in the GDP prompted by an active central bank, be officially recovered.

# 59

# The Unemployment Paradox

Even though contradictory, the fact is that weak sales of the 1930s across multiple sectors occurred during a period with large pockets of increased production. While the high-potential economy obviously underperformed, the impact on the American job market seems excessive. The historic high unemployment rate for the 1930s averaged 18 percent. Despite the economic gains and a military buildup toward the end of the decade, the 1939 unemployment rate only dipped to 17 percent. Compared to past and future panics, the economy's natural tendencies to rebound with rising consumer demand and increasing job opportunities proved especially stubborn.

Hence, the Great Depression becomes the great paradox. How did newfound high productivity and stubborn high unemployment rates coexist during the 1930s?

The first obstacle to analyzing unemployment data is to appreciate that the methods used to establish an accurate unemployment rate has its fair share of idiosyncrasies. For example, who qualifies statistically as employed or unemployed? People who:

- had their weekly fulltime work hours reduced—i.e. to 30 hours, to 20 hours?
- are part of the underground workforce?
- are out of work due to an excessive minimum wage?
- have been *forced* into early retirement?
- earn income through illegal activities?
- abuse entitlements?
- are stay-at-home spouses who would rather work?
- are young adults out of work and living at home?
- rank as long-term unemployed?
- work for a New Deal jobs program?

Fortunately, there is a degree of method amid the madness regarding the accuracy of the unemployment rate. For one, the procedures used to calculate the unemployment rate contain lots of hard data which minimizes its ambiguity and shortcomings. This provides consistency in the methods used to measure unemployment. That consistency from year to year then establishes legitimacy. The Bureau of Labor Statistics (BLS) made slight, inconsequential changes to their statistical methods twice during the Great Depression—in 1930 and 1940 (according to their own notes). A major revision came in 1945. As it turned out, those updated methods correspond to the dramatic economic and social changes that accompanied the new post-war age. In other words, employment statistical methods of the 1920s relate to the 1930s. The big change correlates to 1945 into the 1960s.

Once economists learned to document the finer statistical points of the unemployment rate, policy makers made it their mission to maintain a condition known as full employment. The term is a bit of a misnomer. In modern times during full employment, there are always a few million job seekers out of work (8 million at 5 percent in 2016). To avoid some confusion, the rest of this chapter will refer to the unemployment rate during full employment as the natural rate of unemployment. The United States government has an official natural rate of unemployment goal of three percent. Today, economists consider that milestone optimistic. Instead, they regard five percent as a more realistic goal.

As it turns out, the concept of a natural unemployment rate is a bit elusive serving only as a baseline—a long-term average. In fact, the undulating nature of the business cycle assures that the unemployment rate is in constant flux. The formation of an economic bubble drives the unemployment rate below the standard. A recession sends the unemployment rate above the standard. For the most part, the job market either plays catch-up toward normalcy or swells toward a bubble.

Another statistical quirk worth noting is that the natural unemployment rate does not depend much on the long-term makeup of the economy. Over 100 years of data shows that whether an economy is automated or manual, factory-based or service-based, made up of frugal-government conservatives or big-government spenders, the natural unemployment rate hovers around a tight 4-5% window. The workforce adapts to its environment—another example of where market forces prevail. One of the 1930s progressives' programs that did make a systemic increase in the unemployment rate was the introduction of unemployment insurance—approximately a 1% bump.

If the workforce is so resilient, then what caused the massive unemployment rate spikes during the Great Depression? The key is shock. The American economy is a tough one. It normally rolls along despite immense stress from technological changes, social pressures, etc. Then there are those events that make the economy shudder. Extreme business cycles and big wars are two good examples. The Great Depression had both, along with a third—the rapid increase in the scope and size of an over-bearing federal government.

Previous chapters featured many of the events that made the Great Depression so extraordinary. Figure 59.1 illustrates those moments in graphical form. The solid line charts the Bureau of Labor Statistics (BLS) unemployment rate from 1925 through 1949 and with the dashed line, the estimated natural unemployment rate over the same period. The burst events above the lines represent incidents that lowered the unemployment rate. The burst events below the line represent incidents that elevated the unemployment rate. The illustration exposes the relentless forces that kept employers on edge and millions needlessly unemployed. It also reveals the potency of restrained government intervention. Table 59.1 displays those employment shocks in a table form.

322                      *Lessons Not Learned*

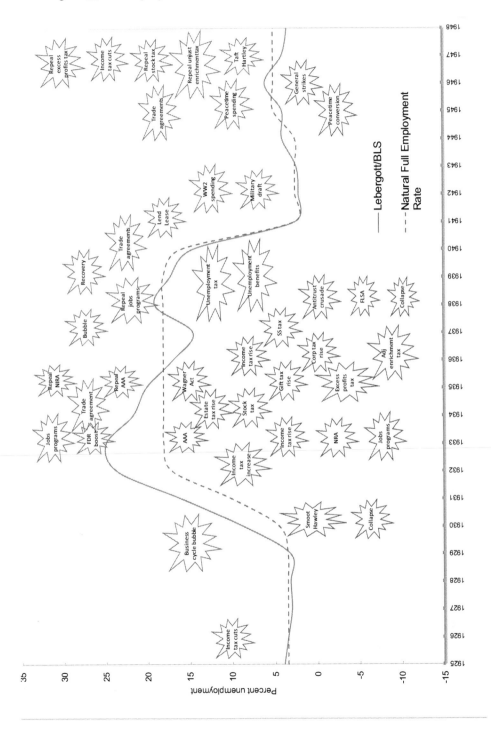

Figure 59.1 Unemployment rate burst events

**Table 59.1** Unemployment rate burst events

| Year | Raised unemployment rate | Lowered unemployment rate |
|---|---|---|
| 1925 | | Income tax cuts |
| 1929 | | Business cycle bubble |
| 1930 | Business cycle collapse | |
| 1930 | Smoot Hawley | |
| 1932 | Income tax increases | |
| 1933 | | Roosevelt election boost |
| 1933 | Income tax increase | |
| 1933 | Stock tax | |
| 1933 | AAA | |
| 1933 | NIRA | |
| 1934 | Jobs programs | Jobs programs |
| 1934 | Increase estate tax | Recip. trade Agreement |
| 1934 | Increase capital stock tax | |
| 1935 | Excessive profits tax | |
| 1935 | Increase gift tax | |
| 1935 | Wagner Act | |
| 1935 | | Repeal AAA |
| 1936 | | Repeal NIRA |
| 1936 | Income tax increases | |
| 1936 | Corp. income tax increase | |
| 1936 | Adjustment enrichment tax | |
| 1937 | | Business cycle bubble |
| 1937 | Unemployment insurance tax | |
| 1937 | Social Security Tax | |
| 1938 | Business cycle collapse | |
| 1938 | | Fair Labor Standard Act |
| 1938 | | Anti-trust crusade |
| 1938 | Collect unemployment insurance | |
| 1939 | | Repeal jobs programs |
| 1940 | | Business cycle Recovery |
| 1941 | | Trade agreements |
| 1942 | | Lend lease |
| 1945 | | WWII buildup |
| 1945 | Peacetime conversion | |
| 1945 | Nationwide general strikes | |
| 1945 | | Trade agreements |
| 1946 | | Peacetime spending |
| 1947 | | Repeal excess profits tax |
| 1947 | | Income tax cuts |
| 1947 | | Repeal capital stock tax |
| 1947 | | Repeal unjust enrichment tax |
| 1947 | | Taft Hartley |

# 60

# Another Great Depression?

Academics study the latter 1920s and 1930s with the primary goal of learning not to repeat the same mistakes that caused the infamous economic panic. After ninety years of reflection, have American leaders developed the tools needed to prevent another Great Depression? Despite analysts' best efforts to avoid past blunders, Mark Twain expressed the proper response years earlier: "History doesn't repeat itself, but it does rhyme." So is the case with the Great Depression. There will never be another similar event, not for lessons learned, but because the extended economic panic occurred during a unique period whose time has passed.

The Great Depression's uniqueness centers on the convergence of two powerful emerging movements. America's factory system turned modern, and the decades-old push to socialize the nation's economy gained significant traction. Since the 1930s, both industrialization and big government have grown by magnitudes. Today, the transformation endures, but their bases are so massive that radical changes close to those that arose nearly a century earlier are unlikely.

Another Great Depression distinction, and one related to the age's modernization, is that, despite the infamous enduring hardships, the nation's economic wellbeing by the end of the recovery far surpassed that of the pre-recession Roaring Twenties. It is doubtful that Americans suffering through future economic catastrophes will be so fortunate.

Although unlikely to repeat the Great Depression, America's future is always in jeopardy, primarily due to the same economic thought that triggered the long recession. In fact, there is a fascinating dichotomy between America's industrial and political-thought paths. When the colonists revolted against England, the motherland was the most advanced society in world history to that point. With little delay, American exceptionalism kicked in and took social and technical progress to a higher level. The two intellectual movements, industry and politics, both advanced into the 1930s. At that point their development diverged underscored by dramatic inflection points. Along the industrial development path new technologies enabled the economy to turn modern as opposed to the ongoing political evolution which favored pessimistic economic models that aligned with oppressing regimes.

The contrast between American industrial growth versus changes in political thought over the last several decades begs for perspective. A comparison between the historic year of 1776 (where the Great Depression saga begins in Chapter 6) to our modern age of the twenty-first century proves telling. The former was a time in which a young America was heavily influenced by both Europe's Industrial Revolution and its great philosophers. How far has America come?

In mid-June 1776, a five-member committee created by the Continental Congress assigned a thirty-two-year-old Virginian, Thomas Jefferson (1743–1826), to write a declaration of independence. A few weeks later, on July 4, 1776, the Continental Congress approved a revised version of Jefferson's draft, titled The Unanimous Declaration of the Thirteen United States of America. Over the ensuing months 56 members of Congress signed the document.

The Declaration of Independence became America's mission statement, emphasized by Jefferson's historic lines,

> We hold these truths to be self-evident, that all men are created equal, that they are endowed by their Creator with certain unalienable Rights, that among these are Life, Liberty and the pursuit of Happiness.

The Declaration, followed by a hard-fought war, paved the way for the United States Constitution. It then took two centuries of dramatic social battles, but Americans were able to sort out most of the archaic traditions and compromises that violated the forefathers' liberty-oriented ideals. As for the nation's economic system, the elegant construction of America's founding documents formed the foundation for a prolific private sector and an impressive network of states, cities, towns, and villages. Today, too many Americans, including academics, tend to take the nation's unique but elegant system for granted. They have traded a craving for freedom and wealth for one of over-tolerance and political correctness and are too willing to accept what the founders would consider a despotic-leaning federal government that promises cradle-to-grave protection.

In 1776, Adam Smith published Wealth of Nations. The renowned economist offered several profound insights into how economies functioned on the eve of the Industrial Revolution. His stance on the power of free choice is still the bedrock of conservative thought, but the one view that best reflects the marvel of a factory-based, free-market economy is his concept of an invisible hand.

The invisible hand increases economic wellbeing by exploiting the market's enormous inertia and unfathomable energy. Modern educators do not teach this economic concept enough. By providing insights into human nature's natural ability to engage producer and consumer without authority they skip the opportunity to impress their students with the wonderful mechanisms that make civilizations prosper.

In technology, James Watt sold his first two steam engines in 1776. The amount of engineering packed into Watt's engine transcended that of any earlier machine. The steam engine added prodigious amounts of horsepower to factories and noticeably sped up travel over land and sea. Watt's work also established the field of engineering and promoted trades critical to the spread of manufacturing.

Today, many thousands of machine types dwarf the utility and sophistication of a Watt steam engine, and there seems to be no technical ceiling as innovation continues at a blistering pace. Unfortunately, a combination of black-box syndrome and academia's apathetic attitude toward the private sector diminishes these accomplishments. Social architects instead look to government to energize technology and subsidize intellectuals in order to fill the void left by a lackluster factory culture, or as they often ingenuously refer to the American economy—post-industrial.

Richard Arkwright's Compton cotton mill began operations in 1776—the world's first large-scale automated factory. Once humanity learned how to convert the short delicate fibers of a cotton ball into durable yarn, all at high speed and with extreme uniformity, they established the ability to revolutionize civilization. Since then, mankind's ingenuity has reinvented the many manufacturing processes countless times, making each generation exponentially wealthier than its predecessor.

Nowadays, in what began with the Luddites, many still judge industrial progress not by the additional tasks performed by labor-saving machines, but instead by the number of jobs destroyed by the machine. Hollywood exacerbates that fear with Frankenstein-type robots destined to annihilate humanity.

Eduard Gibbon (1737–1794), one of the first modern historians, in 1776 published the first of six volumes of his popular The History of the Decline and Fall of the Roman Empire. Gibbon chronicled the struggles of an empire whose wealth towered above every civilization before it and any for the following millennium. The British historian echoed conventional viewpoints which describe the Romans as a population built on advancing technologies, extensive commerce, a world-class military, and slave labor. It took forty decades of Rome's bread and circuses agenda, all paid for by a massive expanding bureaucracy and the demonetization of their coinage to doom the once-mighty empire. Ultimately, they succumbed to simple political greed which created a bloated bureaucracy consumed by hyperinflation.

An academic pioneer, Gibbon wrote history so that others could avoid the mistakes of the past, an effort wasted on much of mainstream academia dedicated to understanding America's economic history. After all, America, with a highly industrialized economy, a liberty-oriented democratic government, and a wealthy populace, all of which are many times more advanced than Rome at its pinnacle, appears to be immune to a similar fate. Nevertheless, not only has academia ignored Roman history, and the many other currency catastrophes over the centuries, they have been able to spin a policy of perennial monetary expansion as an economic positive. Consequently, the technocrats opting for the wizard behind the curtain remain shielded from monetary reality.

✧✧✧

American industry spent the 1920s, 1930s, and 1940s leveraging capitalism's capacity to produce tremendous amounts of egalitarian wealth. The effort not only made the US the world's leading industrial superpower, but that American exceptionalism sparked global industrialization. Alternatively, academics and financiers hindered by their blind spot pursued a simplistic Keynesian economic model that marginalizes the nation's industrial strength, bolsters the benefits of detrimental government, and weakens the dollar. For how long can industry carry the charade?

# Bibliography

1,028 Economists Ask Hoover to Veto Pending Tariff Bill. *New York Times*, 5 May 1930

Alcoa, Our History. *https://www.alcoa.com/global/en/who-we-are/history/default.asp*

Allen, James, and Miller Williams McPherson. *Railroad: Trains and Train People in American Culture*. New York: Random House, 1976

Allen, Robert C. *The British Industrial Revolution in Global Perspective*. Cambridge: Cambridge University Press, 2015

Allen, W. Keith. "Adhesives." Encyclopedia of 20th-Century Technology. Vol. 1. New York, Routledge, 2005

Armagnac, Alden P. "America Starts Work on Biggest Navy." *Popular Science*, January 1941

Anderson, Stuart. "Antibacterial Chemotherapy. "Encyclopedia of 20th-Century Technology. Vol. 1 New York: Routledge, 2005

Anderson, Stuart. "Antibiotics, Development through 1945." Encyclopedia of 20th-Century Technology. Vol. 1 New York: Routledge, 2005

Armagnac, Alden P. "Streamlined Trains Herald in New Speed Era." *Popular Science,* January 1935

Ashton, T. S. *The Industrial Revolution, 1760-1830*. Oxford: Oxford University Press, 1969

Asleep in the Clouds. *Popular Mechanics Magazine,* January 1937

Automatic Drive Car has no Clutch Pedal. *Popular Mechanics Magazine,* December 1939

Avrich, Paul. *The Haymarket Tragedy*. Princeton, NJ: Princeton University Press, 1986

Ballard, Robert D., Ken Marschall, and Patrick Crean. *Exploring the Titanic.* New York: Scholastic/Madison Press Book, 1998

Bauer, Reinhold. "Automobiles." Encyclopedia of 20th-Century Technology. Vol. 1 New York. Routledge, 2005

Baumol, William J., and Alan S. Blinder. *Economics Principles and Policy*. Fort Worth: Dryden Press, 2005

Beacham, Frank. "Chiquola Mill Shootings: The 75th Anniversary." *Frank Beacham's Journal*. September 03, 1990

Beirman, Harold Jr. *The Causes of the 1929 Stock Market Crash*. Westport, CT: Greenwood Press,1998

Benes, James J. "An Industry Evolves: From Lathes to Computers." American Machinist, August 1996

Big Tractors go Diesel. *Popular Mechanics Magazine,* June 1935

Boone, Andrew R. "$2,500,000 Worth of Planes a Day." *Popular Science*, March 1942

Boone, Andrew R. "The Liberator." *Popular Science,* May 1943

Bowen, John T. "Electrical Power in Agriculture." Encyclopedia Britannica. Vol. 7 Chicago: Encyclopedia Britannica, Inc., 1948

Brancker, Air Vice-Marshall Sir William Sefton and Harry P. Kelliher and Daniel Clemens Sayer. "Aviation Civil. " Encyclopedia Britannica. Vol. 1 Chicago: Encyclopedia Britannica, Inc., 1948

Braudel, Fernand. *Civilization and Capitalism, 15th-18th Century*. London: Collins, 1981

Breskin, Charles A. "Plastics." Encyclopedia Britannica. Vol. 18 Chicago: Encyclopedia Britannica, Inc., 1948

Brooks, Peter W. "Aeronautics." A History

of Technology. Vol. V New York: Routledge, 2005

Brown, David J. "Bridges, Long Span and Suspension." Encyclopedia of 20th-Century Technology. Vol. 1 New York: Routledge, 2005

Brunts, Laura, and Theodore Kahn. "The Great Depression." *The Atlantic,* December 2008

Budd, Edward G. "Motor Car." Encyclopedia Britannica. Vol. 15 Chicago: Encyclopedia Britannica, Inc., 1948

Butler, Michael Anthony. *Cautious Visionary: Cordell Hull and Trade Reform, 1933-1937.* Kent, OH: Kent State University Press, 1998

Card, H.S. "Cloud of Fire Guards Drop of Steel." *Popular Mechanics Magazine,* August 1938

Cardwell, D. S. L. *Wheels, Clocks, and Rockets: A History of Technology.* New York: W.W. Norton, 2001

Carrying the Torch for Industry. *Popular Mechanics Magazine,* July 1941

Casey, Robert. *The Model T: A Centennial History.* Baltimore: Johns Hopkins University Press, 2008

Chalmers, Henry. "Tariffs." Encyclopedia Britannica. Vol. 21 Chicago: Encyclopedia Britannica, Inc., 1948

Clark, Mark. "Alloys, Light and Ferrous. "Encyclopedia of 20th-Century Technology. Vol. 1 New York: Routledge, 2005

Cochrane, Charles H. "Freight. "The Encyclopedia Americana. Vol. 1 New York: Americana Corporation, 1946

Colvin, Fred Herbert. "Automatic Machines." Encyclopedia Britannica. Vol. 12 Chicago: Encyclopedia Britannica, Inc., 1948

Colvin, Fred Herbert. "Machine Tools." Encyclopedia Britannica. Vol. 14 Chicago: Encyclopedia Britannica, Inc., 1948

Connelly, T. Garth. *PT Boats in Action.* Carrollton, TX: Squadron/Signal Publications, 1994

Constable, George, and Bob Somerville. A Century of Innovation: Twenty Engineering Achievements That Transformed Our Lives. Washington, DC: Joseph Henry Press, 2003

Corporation Income Tax Brackets and Rates, 1909-2002. Available at: *https://www.irs.gov/pub/irs-soi/02corate.pdf*

Costa, Dora L. "Hours of Work and the Fair Labor Standards Act: A Study of Retail and Wholesale Trade, 1938-1950." NBER Working Paper No. 6855 Issued in December 1998. Available at *https://www.nber.org/papers/w6855*

Cotter, Bill. *1939-1940 New York World's Fair.* Charleston, SC: Arcadia Publishing, 2009

Coulson, Major Thomas. "Research." Encyclopedia Britannica. Vol. 19 Chicago: Encyclopedia Britannica, Inc., 1948

Cousineau, Jennifer. "Laundry Machines and Chemicals. "Encyclopedia of 20th-Century Technology. Vol. 1 New York: Routledge, 2005

Cowan, Ruth Schwartz. *More Work for Mother: The Ironies of Household Technology from the Open Hearth to the Microwave.* New York: Basic Books, 2008 Brooks, Peter W. "Aeronautics." A History of Technology. Vol. V New York: Routledge, 2005

Cowan, Ruth Schwartz. *More Work for Mother: The Ironies of Household Technology from the Open Hearth to the Microwave.* New York: Basic Books, 2008

Cowen, Tyler. *The Great Stagnation: How America Ate All the Low-hanging Fruit of Modern History, Got Sick, and Will (eventually) Feel Better.* New York: Dutton, 2011

Cunningham, William James. "Railways." Encyclopedia Britannica. Vol. 18 Chicago Encyclopedia Britannica, Inc., 1948

de Syon, Guillaume. "Civil Aircraft, Propeller

Drives." Encyclopedia of 20th-Century Technology. Vol. 1 New York: Routledge, 2005

Dehner, Carl. "What Are Plastics Made Of? Popular Science, January 1943

Deutschman, Aaron D. *Machine Design: Theory and Practice*. New York: Macmillan, 1975

Dickinson, H.W. "The Steam-Engine to 1830." A History of Technology. Vol. IV New York: Oxford University Press, 1958

Diesel Limited. *Popular Mechanics Magazine*, October 1939

Donovan, Leo. "Forecast from Detroit." *Popular Mechanics Magazine*, November 1948

Downey, Gregory J. *Technology and Communication in American History*. Washington, DC American Historical Association, 2011

Drover, F.J. "Marine Engineering." Encyclopedia Britannica. Vol. 14 Chicago: Encyclopedia Britannica, Inc., 1948

Easier Shifting in 1938. *Popular Mechanics Magazine*. December 1937

Editors of Consumer Guide. *Classic Ford Trucks*. Lincolnwood, IL: Publishers International, 2012

Elastic Sandwich Cushions New Safety Glass. *Popular Mechanics Magazine*, October 1939

Ellis, C. Hamilton. "The Development of Railway Engineering. "A History of Technology. Vol. V New York: Oxford University Press, 1958

Elton, Sir Arthur. "Gas for Light and Heat. "A History of Technology. Vol. IV New York Oxford University Press, 1958

Faires, Virgil Moring., and Clifford Max. Simmang. *Thermodynamics*. New York: Macmillan, 1978

Falkner, Roland Post. "Machinery and Production." Encyclopedia Britannica. Vol. 14 Chicago: Encyclopedia Britannica, Inc., 1948

Ferguson, Niall. *Civilization: The West and the Rest*. New York: Penguin Books, 2012

Field, Alexander J. *A Great Leap Forward: 1930s Depression and U.S. Economic Growth*. New Haven, CT: Yale University Press, 2012

Field, D.C. "The Internal Combustion Engine." A History of Technology. Vol. V New York Oxford University Press, 1958

Field, D.C. "Mechanical Road-Vehicles. "A History of Technology. Vol. V New York: Oxford University Press, 1958

Fitch, James William. *Motor Truck Engineering Handbook*. Warrendale, PA: Society of Automotive Engineers, 1994

Fleck, Sir Alexander. "Technology and its Social Consequences. "A History of Technology. Vol. V New York: Oxford University Press, 1958

Fleming, Sir Alexander. Encyclopedia Britannica. Vol. 9 Chicago: Encyclopedia Britannica, Inc., 1948

Folsom, Burton W. *New Deal or Raw Deal: How FDRs Economic Legacy Has Damaged America*. New York: Threshold Editions, 2008

Floyd, Barbara L. *The Glass City: Toledo and The Industry That Built It*. University of Michigan Press, 2015

Folsom, Burton W. *The Myth of the Robber Barons*. Hernandon, VA.: Young America's Foundation, 1991

Forbes, R.J. "Petroleum. "A History of Technology. Vol. V New York: Oxford University Press, 1958

*Forming Alcoa Aluminum*. Pittsburgh, PA: Aluminum Company of America, 1962

Francis, Devon. "How New Cars Package People." *Popular Science*, November 1946

Friction Fighters. *Popular Mechanics Magazine*, July 1940

Friedman, Milton, and Anna Jacobson. Schwartz. *A Monetary History of the United States: 1867-1960*. Princeton: Princeton Univ. Press, 1993

Friedman, Milton, and Rose D. Friedman. *Free to Choose: A Personal Statement.* San Diego, CA: Harcourt Brace Jovanovich, 1990

Friedman, Milton. *Capitalism and Freedom.* Chicago: University of Chicago Press, 1982

Friedman, Milton. *Free to Choose.* Free to Choose Enterprise. 1990

Galloway, D.F. "Machine-Tools. "A History of Technology. Vol. V New York: Oxford University Press, 1958

Gertner, Jon. *The Idea Factory: Bell Labs and the Great Age of American Innovation.* London: Penguin Books, 2013

Gibbon, Edward, and Hans-Friedrich Mueller. *The Decline and Fall of the Roman Empire.* New York: Modern Library, 2003

Gifford, Walter Sherman. "Telephone." Encyclopedia Britannica. Vol. 21 Chicago: Encyclopedia Britannica, Inc., 1948

Gilbert, K.R. "Machine-Tools. "A History of Technology. Vol. IV New York: Oxford University Press, 1958

Gilder, George F. *The Spirit of Enterprise.* New York: Harmondsworth Viking, 1985

Glaeser, Martin Gustave. "Public Utilities." Encyclopedia Britannica. Vol. 18 Chicago Encyclopedia Britannica, Inc., 1948

Gordon, John Steele. *An Empire of Wealth: The Epic History of American Economic Power.* New York: Harper Perennial, 2005

Goulden, Joseph C. *The Best Years, 1945-1950.* New York: Atheneum, 1976

Grahame, Arthur. "Why America's Tanks are the World's Best." *Popular Science,* March 1943

Grant, Ellsworth S. *Stanadyne: A History.* Stanadyne, 1985

Grant, Roderick E. "Molding the World of Tomorrow." *Popular Mechanics Magazine,* July 1943

Grant, Roderick E. "Steam versus Diesel." *Popular Mechanics Magazine,* May 1940

Gray, R.B. "Harvesting Machinery." Encyclopedia Britannica. Vol. 11 Chicago: Encyclopedia Britannica, Inc., 1948

Gray, R.B. "Tractors – industrial tractors." Encyclopedia Britannica. Vol. 22 Chicago Encyclopedia Britannica, Inc., 1948

Great Floating Sphere to Be Wonder of Fair. *Popular Mechanics Magazine.* June 1937

Hamor, William Allen. "Research." Encyclopedia Britannica. Vol. 19 Chicago: Encyclopedia Britannica, Inc., 1948

Heckscher, Eli F. *Mercantilism.* London: Routledge, 1994

Heilbroner, Robert L., and William Milberg. *The Making of Economic Society.* Upper Saddle River, NJ: Prentice Hall, 1998

Heldt, Peter Martin. "Tractors." Encyclopedia Britannica. Vol. 22 Chicago: Encyclopedia Britannica, Inc., 1948

Hinds, Fred S. *Industrial Plants.* Philadelphia, PA: G.M.S. Armstrong, 1907

Holbrook, Walter. "Industry's $100,000,000 Tool." *Popular Science,* July 1941

Horner, Fred. "Grinding Machines." Encyclopedia Britannica. Vol. 10 Chicago: Encyclopedia Britannica, Inc., 1948

Horner, Fred. "Machine Tools." Encyclopedia Britannica. Vol. 14 Chicago: Encyclopedia Britannica, Inc., 1948

Horner, Fred. "Milling Machines. "Encyclopedia Britannica. Vol. 15 Chicago: Encyclopedia Britannica, Inc., 1948

Hounshell, David A. *From the American system to mass production, 1800-1932: the development of manufacturing technology in the United States.* Baltimore: Hopkins Univ. Press, 1997

House. Encyclopedia Britannica. Vol. 11 Chicago: Encyclopedia Britannica, Inc., 1948

Howson, Elmer Thomas. "Railways." Encyclopedia Britannica. Vol. 18 Chicago: Encyclopedia Britannica, Inc., 1948

Hurd, Dallas T., and Alden P. Armagnac. "How You'll Use the New Silicones. " *Popular Science,* April 1948

Hyland, P. H., M.E., and J. B. Kommers, M.E. *Machine Design.* New York: McGraw-Hill, 1937

Irwin, Douglas A. *Peddling Protectionism: Smoot-Hawley and the Great Depression.* Princeton, NJ: Princeton University Press, 2011

Jarvis, C. Mackechnie. "The Distribution and Utilization of Electricity. "A History of Technology. Vol. V New York: Oxford University Press, 1958

Jarvis, C. Mackechnie. "The Generation of Electricity. "A History of Technology. Vol. V New York: Oxford University Press, 1958

Johnson, Paul. *A History of the American People.* New York: Harper-Perennial, 1998

Juptner, Joseph P. *U.S. Civil Aircraft Series, Vol. 5.* Blue Ridge Summit, PA: Tab Aero, 1962

Juptner, Joseph P. *U.S. Civil Aircraft Series, Vol. 6.* Blue Ridge Summit, PA: Tab Aero, 1962

Juptner, Joseph P. *U.S. Civil Aircraft Series, Vol. 7.* Blue Ridge Summit, PA: Tab Aero, 1994

Kamm, Lawrence J. *"Designing Cost-efficient Mechanisms: Minimum Constraint Design, Designing with Commercial Components, and Topics in Design Engineering".* Warrendale, PA: Society of Automotive Engineers, 1993

Kaufman, George G. *The U.S. Financial System: Money, Markets, and Institutions.* Englewood Cliffs: Prentice-Hall, 1986

Keefer, Chester Scott, MD. "Sulfonamides." Encyclopedia Britannica. Vol. 21 Chicago: Encyclopedia Britannica, Inc., 1948

Kelly, Barbara M. *Expanding the American Dream Building and Rebuilding Levittown.* Albany, NY: State University of NY, 1993

Kennedy, David M. *Freedom from Fear: The United States, 1928-1945.* New York: Oxford University Press, 1999

Kent. Morgan C. "Auto Transmissions That Shift for Themselves." *Popular Science,* December 1945

Kerr, Charles Jr. "Locomotive." Encyclopedia Britannica. Vol. 14 Chicago: Encyclopedia Britannica, Inc., 1948

Kettering, Charles F., Norman G. Shidle, and Thomas A. Bissell. "Motor Car." Encyclopedia Britannica. Vol. 15Chicago: Encyclopedia Britannica, Inc., 1948

Klein, Maury. *Rainbows End: The Crash of 1929.* Oxford: Oxford University Press, 2003

Kline, Gordon M "Plastics." The Encyclopedia Americana. Vol. 22 New York: Americana Corporation, 1946

Kraehenbueul, John Otto. "Lighting and Artificial Illumination." Encyclopedia Britannica. Vol. 14 Chicago: Encyclopedia Britannica, Inc., 1948

Lacey, Robert. *Ford the Men and the Machine.* Bredbury, England: National Library for the Blind, 1992

Lanchester, Frederick William. "Motor Car." Encyclopedia Britannica. Vol. 15 Chicago Encyclopedia Britannica, Inc., 1948

Law, Alex. "Changing Nature of Work. "Encyclopedia of 20th-Century Technology. Vol. 1New York: Routledge, 2005

Lax, Eric. *The Mold in Dr. Florey's Coat: The Story of the Penicillin Miracle.* New York: Henry Holt, 2005

Leggett, Julian. "Easy Motoring for 1937." *Popular Mechanics Magazine,* January 1937

Leggett, Julian. "Machines That Make Machines." *Popular Mechanics Magazine,* October 1941

Leggett, Julian. "No Blind Driving in 1940s Cars." *Popular Mechanics Magazine,* October 1939

Leggett, Julian. "The 1936 Cars." *Popular Mechanics Magazine,* December 1935

Leggett, Julian. "The Era of Plastics." *Popular Mechanics Magazine,* May 1940

Leggett, Julian. "Wonders of the New York Fair," *Popular Mechanics Magazine,* April 1939

Leggett, Julian. "Making Metal Smooth as Glass." *Popular Mechanics Magazine.* January 1939

Leggett, Julian. Good-By to the Wobble Stick." *Popular Mechanics Magazine.* December1938

Lewis, Hunter. *Where Keynes Went Wrong: and Why World Governments Keep Creating Inflation, Bubbles, and Busts.* Edinburg, VA: Axios Press, 2011

Lieberthal, Edwin M. *Progress Through Precision.* Torrington Company, 1992

Lighter, Sturdier Boxcars Have Plywood Walls. *Popular Mechanics Magazine,* February 1945

Lusk, Harold F. *Business Law: Principles and Cases.* Homewood, IL: R.D. Irwin, 1978

*Machine-Shop Practice.* Scranton, PA: International Textbook Company, 1924

Macintire, Horace James. "Refrigeration and its Application." Encyclopedia Britannica. Vol. 19 Chicago: Encyclopedia Britannica, Inc., 1948

Magee, H.W. "Aluminum the Wonder Metal of the Future." *Popular Mechanics Magazine,* February 1936

Magee, H.W. "Hitch Your Wagon to a Car – Part I." *Popular Mechanics Magazine,* December 1936

Magee, H.W. "Hitch Your Wagon to a Car – Part II." *Popular Mechanics Magazine,* January 1937

Magoun, Alexander B "Audio Recording, Mechanical." Encyclopedia of 20th-Century Technology. Vol. 1New York: Routledge, 2005

Maitland, Lester J. "Aviation Civil." Encyclopedia Britannica. Vol. 2 Chicago: Encyclopedia Britannica, Inc., 1948

Malkin, S. *Grinding Technology: Theory and Applications of Machining with Abrasives.* Chichester: E. Horwood, 1989

Malkin, S. *Grinding Technology: Theory and Applications of Machining with Abrasives.* Chichester: E. Horwood, 1989

Mann, Julia de L. "The Textile Industry Machinery for Cotton, Flax, Wool, 1760-1850." A History of Technology. Vol. IV New York: Oxford University Press, 1958

Margo, Robert A. "Employment and Unemployment in the 1930s. " Journal of Economic Perspective, Volume 7, Number 2—Spring 1993

Margo, Robert A. "Employment and Unemployment in the 1930s. " *Journal of Economic Perspective,* Volume 7, Number 2—Spring 1993. Available at *https://pubs.aeaweb.org/doi/pdfplus/10.1257/jep.7.2.41*

Marsden, Ben. *Watts Perfect Engine: Steam & the Age of Invention.* Cambridge: Icon Books, 2004

Mass, Nathaniel J. *Economic Cycles: An Analysis of Underlying Causes.* Cambridge, MA: Wright- Allen P., 1975

Mayall, William Henry. *Machines and Perception in Industrial Design.* London: Studio Vista, 1968

McConnell, Campbell R. *Economics: Principles, Problems, and Policies.* New York: McGraw-Hill, 1979

McCullough, David G. *1776.* New York: Simon & Schuster Paperbacks, 2006

McElvaine, Robert S. *The Great Depression America, 1929-1941.* New York: Three Rivers Press, 2009

McFadyen, Aubrey D. "Patent No. 2,000,000." *Popular Science,* July 1935

McLachlan, Norman W. "Radio Receivers." Encyclopedia Britannica. Vol. 18 Chicago: Encyclopedia Britannica, Inc., 1948

Meacham, Jon. *Thomas Jefferson: The Art of Power.* New York: Random House Trade Paperbacks, 2013

Meet the Silicone Family. *Popular Mechanics Magazine,* December 1945

Millard, Andre "Loudspeakers and

Earphones. "Encyclopedia of 20th-Century Technology. Vol. 1New York: Routledge, 2005

Mills, Daniel Quinn. *Labor-management Relations.* New York: McGraw-Hill, 1978

Mingos, Howard L. "Transport by Air." Encyclopedia Britannica. Vol. 22 Chicago: Encyclopedia Britannica, Inc., 1948

Modern Miracle. *Popular Mechanics Magazine,* September 1942

Moffatt, L.E. "Household Appliances." Encyclopedia Britannica. Vol. 11 Chicago: Encyclopedia Britannica, Inc., 1948

Moltrecht, K. H. *Machine Shop Practice, Volume 1.* New York: Industrial Press, 1981

Moltrecht, K. H. *Machine Shop Practice, Volume 2.* New York: Industrial Press, 1981

Morgenthau, Henry Jr. "United States of America." Encyclopedia Britannica. Vol. 22 Chicago: Encyclopedia Britannica, Inc., 1948

Morino, Paul. "Unions and Discrimination." *Cato Journal,* Winter 2010

Morris, Charles R. *The Tycoons: How Andrew Carnegie, John D. Rockefeller, Jay Gould, and J.P. Morgan Invented the American Supereconomy.* New York: H. Holt & Co., 2006

Morris, T.N. "Management and Preservation of Food. "A History of Technology. Vol. V New York: Oxford University Press, 1958

Morton, David. "Electronics. "Encyclopedia of 20th-Century Technology. Vol. 1New York Routledge, 2005

Muhlfeld, Joyn E. "Locomotive." Encyclopedia Britannica. Vol. 14 Chicago: Encyclopedia Britannica, Inc., 1948

Mumford, Lewis, et al. *The Myth of the Machine.* New York: Harcourt Brace Jovanovich, 1970

Murphy, Robert P. *The Politically Incorrect Guide to the Great Depression and the New Deal.* Washington, DC: Regnery, 2009

Murtfeldt, E.W. "6,000 Autos Per Day." *Popular Science Magazine,*." October 1940

Murtfeldt, E.W. "Forty Years of the Automobile." *Popular Science Magazine,*." December 1939

Muzzy, David Saville. "United States of America." Encyclopedia Britannica. Vol. 22 Chicago: Encyclopedia Britannica, Inc., 1948

Neumann, Caryn E. "Camera, Polaroid." Encyclopedia of 20th-Century Technology. Vol. 1 New York: Routledge, 2005

New Metals from Powder. *Popular Mechanics Magazine,* March 1940

New York to Europe by Clipper. *Popular Mechanics Magazine,* May 1939

Newcomer, Mabel. "Taxation, Local." Encyclopedia Britannica. Vol. 21 Chicago: Encyclopedia Britannica, Inc., 1948

Nolde, Gilbert C. *All in a Day's Work: Seventy-five Years of Caterpillar.* New York: Forbes Custom Publishing, 2000

O'Neil, William D "Aircraft Design." Encyclopedia of 20th-Century Technology. Vol. 1 New York: Routledge, 2005

O'Brien, Robert. *Machines.* New York: Time, 1964

O'Leary, Michael. *DC-3 and C-47 Gooney Birds.* Osceola, WI, USA: Motorbooks International, 1992

O'Neil, William D. "Aircraft Design. "Encyclopedia of 20th-Century Technology. Vol. 1 New York: Routledge, 2005

Packages Designed to Catch the Eye. *Popular Mechanics Magazine,* April 1941

Parish, William F. "Lubrication." Encyclopedia Britannica. Vol. 14 Chicago: Encyclopedia Britannica, Inc., 1948

Pavelec, Mike. "Aircraft Instrumentation. "Encyclopedia of 20th-Century Technology. Vol. 1New York: Routledge, 2005

Peet, Louise Jenison. "Refrigerators (Household)." Encyclopedia Britannica. Vol. 19

Chicago: Encyclopedia Britannica, Inc., 1948

Perlman, Selig. "Strikes and Walkouts." Encyclopedia Britannica. Vol. 21 Chicago: Encyclopedia Britannica, Inc., 1948

Petroski, Henry. *Invention by Design: How Engineers Get from Thought to Thing.* Cambridge, MA: Harvard University Press, 2004

Post, Robert C. *Technology, Transport, and Travel in American History.* Washington, DC: American Historical Society, 2003

Postwar American Television. Available at: http://www.earlytelevision.org/us_tv_sets.html

Powell, Hickman. "Look Out, Hitler – Here Comes the Flood!" *Popular Science,* May 1943

Powell, Jim. *FDRs Folly: How Roosevelt and His New Deal Prolonged the Great Depression.* New York: Random House, 2005

Pruitt, Bettye Hobbs. *Timken: From Missouri to Mars--a Century of Leadership in Manufacturing.* Boston: Harvard Business School Press, 1998

Radio Robot Warns Pilot of Mountain Ahead. *Popular Mechanics Magazine,* December 1938 Rand, Ayn. *Atlas Shrugged.* New York: New American Library, 1985

Rand, Ayn. *Capitalism: the unknown ideal.* New York: Signet Books, 1967

Read, W.T. " Chemical Industries." Encyclopedia Britannica. Vol. 6 Chicago: Encyclopedia Britannica, Inc., 1948

Redford, Alan H., and Jan Chal. *Design for Assembly: Principles and Practice.* London: McGraw- Hill, 1994

Reid, David T. *Fundamentals of Tool Design.* Dearborn, MI: Society of Manufacturing Engineers, 1991

Reiss, George R. "From Coast to Coast in a Modern Airliner." *Popular Science,* October 1935

Robbins, Lionel. *The Great Depression.* Macmillan and Co. Limited. 1934

Robie T.M. "Diesel Engine. "The Encyclopedia Americana. Vol. 9 New York: Americana Corporation, 1946

Rodengen, Jeffrey L. *The Legend of Ingersoll-Rand.* Ft. Lauderdale, FL: Write Stuff Syndicate, 1995

Romer, Christina. *"Lessons from the Great Depression for Economic Recovery in 2009."* Available at: *https://www.brookings.edu/ wp-content/uploads/2012/04/0309_lessons_ romer.pdf*

Rose, Sharon A., and Neil Schlager. *How Things Are Made.* New York: Black Dog & Leventhal Publishers, 2003

Rosenhain, Walter. "Metallurgy." Encyclopedia Britannica. Vol. 15 Chicago: Encyclopedia Britannica, Inc., 1948

Rouse, Stewart. "What Happens When Kaiser Builds a Ship." *Popular Science,* March 1943

Ruder, W. W. "Tailor-Made Metals." *Popular Mechanics Magazine,* May 1940

S&P Dow Jones Indices. https://us.spindices.com/indices/equity/dow-jones-industrial- average

Salmon, Stuart C. Modern Grinding Process Technology. New York: McGraw-Hill, 1992

Sanders, Godl. V. "Your New Refrigerator." *Popular Science,* October 1945

Sanders, Walter C. "Railways." Encyclopedia Britannica. Vol. 18 Chicago: Encyclopedia Britannica, Inc., 1948

Sandler, B. Z. *Creative Machine Design: Design Innovation and the Right Solutions.* New York: Paragon House Publishers, 1985

Sandler, Martin W. *This Was America.* Media Enterprises, 1980

Sayer, Daniel Clemens. "Hazards of Aviation." Encyclopedia Britannica. Vol. 2 Chicago: Encyclopedia Britannica, Inc., 1948

Schwartz, James E. *Cincinnati Milacron:*

*1884-1984 ; Finding Better Ways*. Cincinnati Milacron, 1984

Schweikart, Larry, and Michael Allen. *A Patriots History of the United States: from Columbus Great Discovery to the War on Terror*. New York: Sentinel, 2007

Service, John A. Glass-making. "The Encyclopedia Americana. Vol. 12 New York: Americana Corporation, 1946

Shlaes, Amity. *The Forgotten Man a New History of the Great Depression*. New York: Harper Perennial, 2008

Shmoop. *History of Labor Unions: Shmoop U.S. History Guide*. Shmoop University Inc.,2010

Shuber, H.R. The Steel Industry. A History of Technology. Vol. V New York: Oxford University Press, 1958

Smith, Edward H. *Mechanical Engineers Reference Book*. Warrendale, PA: Society of Automotive Engineers, 1994

Smith, Jerome F. *The Coming Currency Collapse and What You Can Do about It*. New York: Books in Focus, 1980

Smith, Robert H. *Text-book of Advanced Machine Work*. Boston: Industrial Education Book Company, 1917

Solberg, Harry L., Orville C. Cromer, and Albert R. Spalding. *Thermal Engineering*. New York: John Wiley & Sons, 1960

Sorenson, Lorin. *The Ford Road*. St. Helena, California: Silverado Publishing Company, 1978

Sowell, Thomas. *Applied Economics: Thinking beyond Stage One*. New York: Basic Books, 2004

Sowell, Thomas. *Black Rednecks and White Liberals*. San Francisco, CA: Encounter Books, 2006

Sprovieri, John. "A Century of GE Appliance Manufacturing." *Assembly Magazine*, March 29, 2017

Steel Demand Reflects New Living Standards. *Popular Mechanics Magazine*. August 1935

Stowers, A. The Stationary Steam-Engine. A History of Technology. Vol. V New York: Oxford University Press, 1958

Strawser, Cornelia J. *Business Statistics of the United States: Patterns of Economic Change*. 14th Edition. Lanham, MD. Bernan Press. 2009

Stuffing Box. Encyclopedia Britannica. Vol. 21 Chicago: Encyclopedia Britannica, Inc., 1948

Tarbell, Ida M. *The History of the Standard Oil Company*. New York: McClure, Phillips & Co., 1904

Taussig, Frank William. "Tariffs." Encyclopedia Britannica. Vol. 21 Chicago: Encyclopedia Britannica, Inc., 1948

Taylor, Lloyd William. "Telephone." Encyclopedia Britannica. Vol. 21 Chicago: Encyclopedia Britannica, Inc., 1948

Teague, Walter Darwin. Planning the World of Tomorrow." *Popular Mechanics Magazine* December 1940

Teale, Edwin. "Hauling Stunts by Motor Trucks." *Popular Science,* January 1935

Teale, Edwin. "Diesel Engines Usher in a New Age of Power." *Popular Science,* October 1935

Teale, Edwin. "Fifty Years of Aluminum." *Popular Science,* February 1936

Teale, Edwin. "Plastic in the War." *Popular Science,* June 1941

Teale, Edwin. "Building a World's Fair." *Popular Science Magazine*. March 1938

Thalenfeld, L. "You're Going to Freeze." *Popular Science,* February 1947

The "American System" of Manufacturing. *Compressed Air Magazine,* September 1988

The 1935 Cars. *Popular Science Magazine*. *February 1935*

The Advance of Diesel. *Popular Mechanics Magazine,* June 1938

The All-Electric Farm. *Popular Mechanics Magazine,* September 1939

The Amazing Story of Stainless Steel. Popular Mechanics Magazine, July 1936

The Amazing Story of the Automobile. *Popular Mechanics Magazine.* December 1936

The American System of Manufacturing. *Compressed Air Magazine.* September 1988

*The Ford Book of Styling.* Dearborn: Ford Motor Company, 1963

The Inside Story. *Popular Science,* July 1948

The Machine Tool Hall of Fame. *Tools & Technology,* Fall 1998

The Most Powerful Diesel on Wheels. *Popular Mechanics Magazine,* June 1941

The New Era of Railroading, Part I. *Popular Mechanics Magazine.* October 1936

The New Era of Railroading, Part II. *Popular Mechanics Magazine.* November 1936

The Romance of the Locomotive. *Popular Mechanics Magazine,* October 1938

The World of Tomorrow. *Popular Mechanics Magazine,* August 1938

They Move Earth. *Popular Mechanics Magazine,* April 1945

Thum, Earnest Edgar." Vanadium Steel" Encyclopedia Britannica. Vol. 22 Chicago Encyclopedia Britannica, Inc., 1948

Toynbee, Arnold. *The Industrial Revolution. With a Pref. by Arnold J. Toynbee.* Boston: Beacon Press, 1962

Turn it with Carbide Tools. *Popular Mechanics Magazine.* February 1948

Turner, Frederick Jackson. "United States of America." Encyclopedia Britannica. Vol 22 Chicago: Encyclopedia Britannica, Inc., 1948

U.S. Bureau of the Census, comp. *Historical Statistics of the United States, 1789-1945.* Washington, DC: 1949

U.S. Bureau of the Census, Continuation to 1952 of *Historical Statistics of the United States, 1789-1945.* Washington, DC: 1954

United States Federal State and Local Government Spending Fiscal Year 1928 (-1949) in $ billion. available at: *https://www.usgovernmentspending.com/year_spending_1928 (-1949) USbt_19bs2n#usgs302*

Van Duyne, Schuyler. "Giant Locomotive." *Popular Science,* May 1941

Van Duyne, Schuyler. "Talking Train Tours Fair Exhibit." *Popular Science,* July 1939

Wallace, H. Paull and James W. Scade. "Tyre." Encyclopedia Britannica. Vol. 22 Chicago: Encyclopedia Britannica, Inc., 1948

Waller, Spencer Weber, The Antitrust Legacy of Thurman Arnold. St. John's Law Review, Vol. 78, No. 3, pp. 569-613, Summer 2004. Available at SSRN: *https://ssrn.com/abstract=648203*

Walsh, Ronald A. *McGraw-Hill Machining and Metalworking Handbook.* New York: McGraw-Hill, 1994

Waltz, George H., Jr. "How Good are Synthetic Tires?" *Popular Science,* December 1946

Washington, Booker T. "The Negro and the Labor Unions." The Atlantic Monthly, June 1913

Weber, Austin. "A Historical Perspective." *Assembly Magazine,* August 19, 2001

Weber, Austin. "Boeing's Innovative Aircraft Changed Aviation and Manufacturing History." *Assembly Magazine,* April 20, 2016

Weber, Austin. "General Electric Leads the Way in Locomotive Manufacturing." *Assembly Magazine,* March 29, 2017

Weber, Austin. "GM Centennial: 100 Years of Manufacturing Milestones." *Assembly Magazine,* July 2, 2008

Weber, Austin. "Line Shafts and Belts." *Assembly Magazine,* October 1, 2003

Weber, Austin. "Rivets, Steam and Sweat." *Assembly Magazine,* January 5, 2005

Weber, Austin. "The Moving Assembly Line Turns 100." *Assembly Magazine,* October 1, 2013

Weber, Austin. "The Rouge an Industrial Icon." *Assembly Magazine,* May 20, 2003

Weber, Austin. "Then and Now: The Man Behind the Moving Assembly Line." *Assembly Magazine,* February 1, 2003

Weber, Austin. "Whirlpool Centennial: From Humble Roots to Global Production Power." *Assembly Magazine,* November 2011

Weck, Manfred. *Handbook of Machine Tools.* Chichester: Wiley, 1984

Welding for War and Peace. *Popular Mechanics Magazine,* June 1943

Westwood, John, and Ian Wood. *The Historical Atlas of North American Railroads.* New York: Chartwell Books, 2011

White, George S. *Memoir of Samuel Slater, the Father of American Manufactures.* Philadelphia, 1836

Williams, Walter E. "The Welfare State's Legacy." September 20, 2017. Available at *http://walterewilliams.com/the-welfare-states-legacy/*

Wilson, Charles Morrow. "Fire Under Your Hood." Popular Mechanics Magazine, July 1938

Wilson, Harold Albert. "Electricity." Encyclopedia Britannica. Vol. 7 Chicago: Encyclopedia Britannica, Inc., 1948

Wittaker, Wayne. "Debut of the '49 Ford." *Popular Mechanics Magazine,* July 1948

Wonder Drug Made From Mold Aids Wounded. *Popular Mechanics Magazine,* November 1943

Woods, Thomas E. *The Politically Incorrect Guide to American History.* Washington, DC: Regnery Publishing, 2004

Worthington, William E. Jr. "Construction Equipment. "Encyclopedia of 20$^{th}$-Century Technology. Vol. 1 New York: Routledge, 2005

Worthington, William E. Jr. "Power Tools and Hand Held Tools "Encyclopedia of 20$^{th}$ Century Technology. Vol. 2 New York: Routledge, 2005

Wright, Charles." Locomotive. "Encyclopedia Britannica. Vol. 14 Chicago: Encyclopedia Britannica, Inc., 1948

Young, Ronald. "Communications. "Encyclopedia of 20th-Century Technology. Vol. 1 New York: Routledge, 2005

Zworykin, Vladimir Kosmiand and George Ashman Morton. "Television." Encyclopedia Britannica. Vol. 21 Chicago: Encyclopedia Britannica, Inc., 1948

# Index

**A**

Abbott Laboratories 259
AC Spark Plug 71
Adams, John Quincy 110
ad-valorem tariff 126, 128
African American 7, 8, 148, 254, 286
Agricultural Adjustment Act 135, 168, 241
agriculture 44, 53, 77, 148, 160, 164, 165, 242, 268
air conditioning 37, 221, 222, 264, 275
aircraft 35, 73, 74, 75, 76, 81, 91, 197, 198, 199, 200, 206, 232, 233, 261, 277, 278, 283
aircraft, civilian
    clippers 199
    Douglas DC3 ix, 197, 198, 199, 335
    Ford Trimotor 75, 76, 197, 198
    Wright airplane 74
aircraft, military
    B-17
        232
    B-24
        231, 232
    B-25
        232
    B-29
        232
    Curtiss JN-4 75
        75
    Douglas C-47 199, 335
airlines 11
    American Airlines 75, 198
    Pan American Airways (Pan Am) 75, 199
    TWA 197, 198

airports 11
Alcoa 175, 176, 177, 178, 328, 331
alternating current (AC) 31, 84, 85
aluminum 11, 63, 75, 91, 176, 177, 178, 184, 194, 197, 219, 233, 244, 264
Amalgamated Association of Iron and Steel Workers 145
Ambassador Bridge 10
American exceptionalism 41, 42, 99, 116, 175, 234, 253
A.M. Freeland 53
ammonia 222
AM radio 9, 81, 275
antitrust 115, 116, 117, 168, 173, 174, 175, 178, 259
appliances 10, 58, 63, 83, 88, 117, 209, 253, 291, 293, 294, 312
Arkwright, Richard 47, 48
Armstrong, Edwin 276
Atlas Shrugged 211, 336
AT&T 37, 276, 282
automobile 6, 10, 11, 12, 19, 24, 33, 35, 36, 38, 65, 66, 67, 68, 70, 71, 77, 90, 103, 104, 113, 115, 158, 167, 174, 186, 187, 188, 190, 191, 192, 212, 213, 247, 261, 265, 269, 273, 275, 284, 285, 287, 300, 302, 310, 312
    Cadillac 67, 71
    Chevrolet 71, 187, 273
    Chrysler 10, 72, 104, 191, 285
    Dodge 33, 67, 189, 190
    Ford Model T 67, 68, 69, 70, 72, 78, 90, 95, 103, 167, 221, 273, 310, 329, 330
    Oldsmobile 71
    Packard 70, 186, 193

Pontiac  35, 71
Studebaker  273

# B
Baekeland, John  224
Baltimore & Ohio Railroad  54
bank failures  14, 25, 29, 128, 139
Barrow, Rudolf  207
Battle of Blair Mountain  147
bearing
 ball bearing  66, 87, 88
 tapered roller  88, 89, 345
Bear Mountain Bridge  10
Beirman, Harold Jr.  118
Bell Labratories  283, 284
Bell Telephone  11–348
Bernays, Edward  33
Bessemer, Henry  31, 31–348, 58, 211
Bessemer process  31, 58, 177, 211, 212
Best Friend of Charleston  54
bicycle  66, 71, 73, 105
Birdseye, Clarence  293
Black Tuesday  12, 274
bottling  201
box making  204
Brandeis, Louis  241
Bretton Woods, New Hampshire  307
Bridgeport  207, 208, 209
Bridgeport milling machine  207, 208
Brooklyn Bridge  10
Brown and Almy  48, 49
Brown, Moses  48
Brown & Sharpe  53
Bureau of Labor Statistics  18
Bush, George W.  238
business cycle  20, 26, 64, 101, 102, 315

# C
camera  80, 277
capitalism  1, 2, 3, 4, 25, 40, 41, 63, 98, 99, 100, 103, 104, 106, 107, 108, 109, 128, 162, 163, 166, 167, 171, 179, 182, 229, 250, 299, 302
Capone, Al  9
carbon black  104, 299
Carnegie, Andrew  31, 58, 60, 211, 212, 335
Carnegie Steel Company  145, 211
Carrier Corporation  223
Carrier, Willis  221
Caterpillar Tractor Company  268
celluloid  224
centerless grinder  91
Chain, Ernst B.  259
Chaplin, Charlie  8
Charles Pfizer  259
Chevrolet, Louis  71
Christensen, Niels  277
Chrysler Building  10
Cincinnati Milling Machine  91, 246
Civilian Conservation Corps (CCC)  16
Civil War  1, 53, 56, 58, 211, 292
clothed in the public interest  118
cobalt-HSS  214
Coconut Grove fire in Boston  259
Cohen, Ben  241
communism  107, 121, 241
Con Edison  37
Coolidge, Calvin  122, 123
Cooper, Peter  54
Corcoran, Tommy  241
Cordell Hull  236, 329
Corliss, George Henry  53
Corliss steam engine  31
Costa, Dora L.  154
cotton  44, 47, 48, 49, 50, 53, 57, 108, 151, 160, 161, 164, 181, 221, 295, 297, 334

cotton gin 50
Cowen, Tyler 40
Crystal Palace 31
Cummins, Clessie 193
Cummins Engine Company 193

## D

Darby, Michael R. 18
Darrow, Clarence 171
D-Day invasion 259
Deere, John 77
deflation 124, 126, 128, 137, 148
Delaware River Bridge 10
Democratic Party 2
Dempsey, Jack 8
Department of Agriculture (DOA) 135, 163, 165, 249, 258, 278
Department of Commerce 122, 308
Department of Justice (DOJ) 115, 172, 173, 234, 258
  Antitrust Division 173, 175, 177
de Tocqueville, Alexis 40
diesel-electric locomotive 36, 264, 266, 314
Diesel, Rudolf 32, 192
direct-current (DC). See also DC
Divco 195
Di Vinci, Leonardo 87
Douglas Aircraft Company 198
Douglass, Frederick 170
Dow Chemical Company 224
Dow Jones Industrial Average (DJIA) 7
drill press 52
Durant, Billy 15, 70, 71, 221
Dust Bowl 15, 160, 161
dynamo 87

## E

Eastman Kodak Co. 279
Edison, Thomas 31, 66, 83, 84, 85, 177, 181, 182, 262, 277, 280, 301
Edward G. Budd Company 264
Eiffel Tower 31
Eighteenth Amendment 9
Einstein, Albert 36
electrical power grid 10
electric factory 92, 93, 96
electric lighting 10
electric motor 12, 68, 87, 88, 92, 95, 96, 177, 207, 262, 269, 272, 295
electric-powered irrigation 10
electronics 11
Elektro 37
elevator 31, 53
Empire State Building 10
equation of exchange 307, 308

## F

facsimile machine (fax) 36
Fair Standards Labor Act of 1938 172
Faraday, Michael 87
farm 10, 12, 21, 25, 28, 29, 33, 38, 51, 59, 66, 77, 78, 79, 104, 111, 120, 121, 123, 128, 160, 161, 162, 163, 164, 165, 167, 191, 202, 241, 254, 271, 272, 311
Farm Security Administration viii, ix, 16, 17, 191, 201, 204, 268
Federal Aviation Administration (FAA) 249
Federal Communications Commission 113, 249, 276
Federal Deposit Insurance Corporation (FDIC) 133
Federal Reserve Act of 1913 138
Federal Reserve System (Fed) 4, 7, 28, 100, 137, 138, 140, 178, 309, 315, 317

fertilizer 164
fiat money 2, 141
Field, Alexander J. 40
fifth wheel 193
First 100 Days 21
Fischer, Friedrich 88
Fischers Aktein-Gesellschaft (FAG) 88
Fitzgerald, F. Scott 33
flappers 9
Fleming, Alexander 256
Florey, Dr. Howard W. 256, 257, 259, 260, 333
FM radio 36, 276
Fokker 197, 198
Ford, Edsel 78
Ford, Henry x, 59, 66, 67, 68, 69, 70, 71, 72, 75, 78, 94, 95, 103, 104, 125, 167, 169, 170, 271, 273, 274, 275
Ford Motor Company 67, 69, 70, 72, 78, 169, 232, 337
Fordney-McCumber Tariff 111
forklift 269, 270, 313
Fourteenth Amendment 254
Franklin, Benjamin 83
free trade 125, 236, 307
Freon 221, 222, 223, 292
Frick, Henry 145, 146
Friedman, Milton 27, 110, 300
Frigidaire Company 221
Fulton, Robert 54
fur trade 44
Futurama 34
F. W. Woolworth Company 59

## G

General Electric (GE) 33, 82, 85, 183, 224, 264, 266, 292. See also GE
General Foods 294
General Motors ix, 15, 34, 35, 70, 71, 104, 173, 188, 190, 191, 192, 193, 221, 222, 264, 266, 273, 284
generator 31, 79, 85, 92, 217, 218, 264
George Washington Bridge 10
Gibbon, Eduard 326
Gilded Age 58, 59, 60, 143, 181, 211
gold standard 2
Gore, Al 127
Gould, Jay 145, 335
Grassley, Chuck 272
Greenback Party 98
Greenspan, Alan 178
grinding wheel 91
gross domestic product (GDP) 1, 19, 19–348, 307, 309, 313. GDP GDP
Gutenberg, Johann 80
gypsum wallboard 289, 290

## H

Hall, Charles M. 176
Hall-Héroult 176
Hall, John Hancock 51
Hamilton, Alexander 50, 109, 137
Hand, Learned 178
Harding, Warren G. 110, 122, 131
Harlem Renaissance 8
Hartley, Fred A. Jr. 247
Harvard University 130, 335
Hawley, Willis C. 124. Mr. Hawley GDP
Heatley, Dr. Norman 257
Héroult, Paul 176
highway 11, 34, 275
    Lincoln Highway 11
    Taconic State Parkway 11
    US 1 11
Hitler, Adolf 230
Hollywood 8

Holt Manufacturing Company  268
Hoover, Herbert  21, 27, 28, 41, 99, 109, 120, 121, 122, 123, 124, 125, 126, 130, 131, 132, 134, 135, 162, 163, 164, 236, 239, 250, 328
Hoover, Lou  120
Hoover War Library  121
Horwitz, Steven  135
hospitals  12
Hume, David  108
Hunt, Mary  258
Hyatt, John Wesley  224
hydraulic control  268
hydraulic power  275
hydroelectric  85, 183

I
immigration  7, 149
incandescent light bulb  81, 83, 177, 182, 184, 215
Industrial Revolution  47, 48, 52, 53, 56, 58, 59, 77, 105, 109, 181, 182, 190, 206, 211, 267, 295, 328
infant industries  110
inflation  70, 137, 140, 154, 167, 220, 308, 309, 310
Ingersoll Rand  193, 264
interchangeable parts  51, 53
internal combustion engine  6, 29, 32, 41, 65, 66, 74, 78, 192, 207, 261, 313
International Business Machines (IBM)  36. See also IBM
International Harvester  ix, 79, 191, 271, 293
International Monetary Fund (IMF)  307
Interstate Commerce Commission (ICC)  111

iron  44, 45, 54, 55, 58, 69, 77, 78, 86, 91, 112, 113, 211, 213, 217, 218, 244, 271, 280

J
Jefferson, Thomas  334
Johnson, Hugh S.  168
Johnson Lyndon B.  238
Johnson, Paul  27, 63
Jolson, Al  9
Jones, Bobby  8

K
Kaiser, Henry J.  232
Kelvinator Company  221
Kennametal  215
kerosene  58, 83, 114, 115, 177, 183, 184
Kettering, Charles F.  71
Keynesian economics  4, 98, 302
Keynes, John Maynard  4
Kinetic Corporation  224
Kitty Hawk, North Carolina  73, 74
Knudsen, Bill  231

L
Labor Management Relations Act of 1947  248
labor strikes  24, 238, 273
  Battle of Blair Mountain  147
  Great Railroad Strike of 1877  143
  Haymarket Affair  144
  Herrin Massacre  147
  Homestead Strike  145
  Lattimer Massacre  146
  Ludlow Massacre  146, 147
LaGuardia Airport,  33
La Guardia, Fiorello  32
Lamont, Thomas  125
Land Edwin  277

Lange, Dorothea 17
Lend-Lease Policy. 231
Levitt, Abraham 286
Levittown 254, 286, 333
Liberty ships 232
Lincoln, Abraham 130
Lincoln Electric 218, 219
Lincoln Highway 11
line shaft 94
Locke, John 108
Los Angeles Coliseum 8
lubrication 206, 263

## M
magneto 69
Manhattan Project 278
Marconi, Guglielmo 80
Marx, Groucho 15
Mazda lamp 183, 184
McCormick, Cyrus 53, 77, 181
McKenna, Philip M. 215
McLaughlin, "Sam" R. S. 70
mechanical reaper 77
medicine 12, 253, 257
mercantilism 107, 109, 332
Michelin 104
Midgley, Thomas, Jr. 222
milling machine 52, 207, 208, 300
minimum wage 153, 154, 240, 241, 319
Moley, Raymond 135
monetary velocity 307
money supply 18, 19, 21, 26, 28, 137, 139, 307, 311
Monsanto Chemical Co. 279
Morgan, J.P. 31, 58, 85, 125, 183, 211, 335
Morrill Tariff of 1861 110
Morse, Samuel 53, 80
Moses, Robert 33

motorized factory 6, 29, 41, 103, 313
music 8, 9, 16, 82, 158, 276

## N
National Broadcasting Company (NBC) 36. See also NBC; See also NBC
National Industrial Recovery Act (NIRA) 134, 150, 153, 168, 241
National Labor Relations Act 152, 172
National Labor Relations Board (NLRB) 152, 249
National Recovery Administration (NRA) 135, 168, 169, 170, 171, 172, 175, 240, 241
  Codes of Fair Competition 169
New Deal 1, 4, 16, 18, 19, 20, 21, 23, 25, 25–348, 27, 28, 32, 40, 63, 99, 117, 127, 128, 129, 130, 133, 134, 135, 136, 137, 138, 140, 143, 153, 157, 160, 163, 164, 165, 168, 171, 172, 173, 179, 205, 229, 233, 234, 235, 238, 239, 240, 241, 243, 247, 248, 249, 250, 308, 313, 319, 331, 335
New York City 10, 11, 29, 32, 33, 48, 49, 53, 54, 82, 83, 141, 145, 157, 225, 279, 282, 283
New York World's Fair 1939-1940 32, 33, 34, 37, 38, 39, 40, 269
Nineteenth Amendment 9
Nixon, Richard 238
Nobel Prize 260, 282, 283, 285
North American Free Trade Agreement (NAFTA) 127
nylon 37, 224, 225, 226, 244, 296, 314

## O
Obama, Barack 305
Organized Labor 247
O-ring 277
Otis, Elisha 31, 53, 181
Otto, Nikolaus 65

Owens Bottle Machine Company 203
Owens-Corning Fiberglass Company 224
Owens, Michael 203
Oxford University's Center of Pathology 257

P
packaging 38, 173, 201, 204, 205, 213, 226, 300, 302
pencil 299, 300, 301, 302
Perisphere 38
Perkins, Francis 133, 150
Perot, Ross 127
Philco 285, 292
phonograph 9
Pinkerton 145
plastics 36, 224, 225, 226, 227, 253, 290
   Bakelite 224
   celluloid 224
   nylon 224
   Plexiglas 35, 36, 232
   polyethylene 36
   polyurethane 36
   PVC 36, 226
   Rayon 167
   silicone 224
plywood 38, 288, 289
Polaroid 277, 335
Postum Cereal Company (Post) 294
Powell, Jim 140
Pratt & Whitney 53
printing press 53, 80, 140
Procter & Gamble 33
progressives 2, 3, 41, 42, 98, 105, 106, 107, 112, 153, 162, 164, 166, 168, 179, 235, 236, 318
Prohibition 9
protectionism 109, 332

Public Works Administration (PWA) 135
punched card 36

R
RADAR 278
radio 6, 9, 11, 21, 36, 80, 81, 82, 132, 200, 242, 275, 276, 278, 281, 285
   AM 11
   short-wave 11
Radio Corporation of America (RCA) 36. *See also* RCA
Rand, Ayn 211
Raytheon 285
Read, Leonard E. 299, 300, 301, 336
Reciprocal Trade Agreement Act 235
refrigeration 161, 165, 195, 221, 222, 291, 293, 294
refrigerator x, 221, 222, 223, 291, 292, 293, 295, 296
Republican Party 2
RMS Titanic 81
Roaring Twenties 3, 4, 6, 7, 12, 14, 14–348, 16, 25, 29, 63, 113, 179
Robbins & Lawrence 53
Robbins, Lionel 15
Roberts, Owen 241
Robinson, Jackie 254
Rockefeller, John D. 58, 114, 115, 183, 335
Rockne, Knute 8
Romer, Christina 318, 336
Roosevelt, Eleanor 130
Roosevelt, Franklin D. 21, 28, 134, 150, 232, 238
Roosevelt, Theodore "Teddy" 130
Rose Bowl 8
Rousseau, Margaret Hutchinson 259
Ruth, Babe 8

## S

Samuelson, Paul  23, 238, 253
Schechter brothers  240, 243
Schwartz, Anna  27
Sears & Roebuck  10, 59
Securities and Exchange Commission (SEC)  249
sewing machine  55, 86, 105
Sherman Act  115, 117, 173, 175, 177, 178, 258
Shewhart, Walter  283
ships  206
Shlaes, Amity  27, 240
Shockley, William  284
Silicon Valley  11, 285
Slater, Samuel  48, 49
slavery  56, 57, 148
Sloan, Alfred B.  71
Smith, Adam  109, 124, 237, 300
Smoot-Hawley Tariff Act  27, 124, 127, 235
Smoot, Reed  124
socialism  121, 134, 235, 241
Social Security Administration  249
Soldier Field  8
Sony  285
Sorenson, Charlie  69
Southern Illinois Coal Company  147
Soviet Union  22, 98, 122, 171, 206, 231
Space Needle  31
Standard Oil  58, 59, 114, 115, 116, 117, 177, 183, 337
Stanley Lebergott  18
steam engine  10, 31, 36, 52, 53, 54, 86, 262, 265, 266
steam turbine  86, 92
steel  10, 11, 31, 35, 36, 38, 58, 66, 68, 75, 77, 78, 79, 87, 88, 105, 112, 145, 146, 167, 176, 177, 188, 195, 202, 211, 212, 213, 214, 215, 216, 217, 218, 219, 233, 244, 247, 264, 268, 272, 290, 292, 296
Steinbeck, John  16
Stein, Ben  127
St. Mary's Hospital  256
stock market crash-1929  7, 17, 98, 125, 128, 141, 151, 167
Strutt, Jedediah  48
Swan, Joseph  182

## T

Taconic State Parkway  11
Taft-Hartley Act  248
Taft, Howard  123
Taft, Robert A.  247
taxes  19, 23, 27, 29, 98, 111, 128, 134, 135, 158, 162, 165, 238, 250
telegraph  6, 53, 282
telephone  11
television  27, 35, 36, 37, 63, 254, 276, 283, 300
Tennessee Valley Authority (TVA)  134, 157, 249
Tesla, Nicola  31, 84, 85
Texas Instruments  285
The American System of Manufacturing  51, 337
thermionic valve (vacuum tube)  81
Thirteenth Amendment  148
Thom, Dr. Charles  258
Time Magazine  122, 169
Timken Company  89
Timken, Henry  88, 89
tires  66, 68, 104, 134, 191, 194, 226, 244, 265, 269, 272, 299
Torrington Company  105, 333
Town of Tomorrow  37

tractor viii, ix, 77, 78, 79, 104, 190, 191, 193, 194, 268, 269, 270, 271, 272
transistor 254, 284, 285
Triborough Bridge 33, 157
truck 71, 105, 158, 190, 191, 192, 193, 194, 195, 199, 288
  Diamond T 191
  Kenworth 190, 191, 194
  Mack 105, 190, 191, 194
  REO 191
Trylon 38
Tugwell, Rex 135
tungsten carbide 215, 216
tungsten filament 183, 184
Twain, Mark 58
Twenty-First Amendment 9

U
unemployment rate 1
United Mine Workers of America 146, 147
United Nations Monetary and Financial Conference 307
United States Department of Agriculture (USDA) 162, 278
United States Postal Service 59
United Steel 38
United Textile Workers 151
University of California 278
US Census 139
US Census Bureau 139, 167
US Constitution 56, 99, 243
US Department of Transportation 275
US Navy 31, 131, 232, 283, 289, 328
US Supreme Court 116
US Treasury 2, 21

V

Vanderbilt, Cornelius 54, 57
Victory ship 232
vitaphone 8, 9, 283
Voder 37
Voltaire 108

W
Wallace, Henry A. 163
Wall Street 12, 14, 15, 19, 133, 235
War Industrial Board 168
Warner Brothers 8
Warner & Whitney 53
war of the currents 84, 85
Washington, Booker T. 170
Watt, James 53, 54, 181
Wealth of Nations 109, 125
welding 38, 188, 217, 218, 219, 266, 293
  Submerged arc welding 219
  tungsten inert gas 219
Wendy the Welder 219
Western Electric 11
Westinghouse Electric 31, 37, 84, 85, 86, 177, 293
Westinghouse, George 31, 84
Whalen, Grover 32
Whitney, Eli 50, 51, 181
Williamsburg Bridge 10
Williams, Walter E. 170
Wilson, Woodrow 110, 121, 131
Wolf, Fred 221
Works Progress Administration (WPA) 16, 157
work stoppages 156
World War I 1, 7, 11, 16, 26, 74, 75, 78, 81, 98, 100, 104, 117, 121, 131, 141, 153, 160, 198, 201, 202, 217, 231, 259, 268, 288
World War II 1, 3, 16, 19, 22, 23, 25, 27, 28, 29, 30, 32, 34, 40, 41, 94, 101, 135, 142,

159, 219, 225, 229, 232, 233, 236, 238, 243,
244, 253, 256, 263, 265, 278, 283, 287,
289, 290, 294, 295, 301, 308, 312, 313
Wright Brothers  6, 73
  Wright, Orville  73, 74
  Wright, Wilbur  73, 74
W. T. Grant  59

## Y
Yamamoto, Isoroku  231
Yankee Stadium  8

# About the Author

Jim's career spanned 40 years as a manufacturing engineer with experience in the automotive, aerospace, power transmission, and other industries. His primary expertise is in precision machine design, but he has had projects analyzing the cost and efficient flow of goods across the factory floor. As a result, Jim understands the advantages and challenges of mass production along with insights into its modernization processes. His first book is *Blind Spot: How Industry Rescued America's Great Depression Economy*.

For questions or comments please contact jim at jsaundbooks@gmail.com

Follow Jim on  @jimsaunders76

Made in the USA
Middletown, DE
12 October 2022

12575982R00201